Systems & Control: Foundations & Applications

M. Bala Subrahmanyam

Finite Horizon H_∞ and Related Control Problems

Birkhäuser
Boston • Basel • Berlin

M. Bala Subrahmanyam
Flight Dynamics and Control Branch
Air Vehicle & Crew Systems Technology Dept.
Naval Warfare Center
Aircraft Division
Warminster, PA 18974-0591

Library of Congress Cataloging In-Publication Data

Subrahmanyam, M. B. (M. Bala), 1949-
Finite horizon H [infinity] and related control problems / M. B. Subrahmanyam.
p. cm. -- (Systems & control)
On t.p. "[infinity]" appears as the infinity symbol.
Includes bibliographical references and index.
ISBN-13:978-1-4612-8718-6
1. H [infinity symbol] control. 2. State-space methods.
I. Title. II. Series.
QA402.3.S88 1995 94-46728
629.8'312--dc20 CIP

Printed on acid-free paper

Birkhäuser
Softcover reprint of the hardcover 1st edition 1995

ISBN-13:978-1-4612-8718-6 e-ISBN-13:978-1-4612-4272-7
DOI: 10.1007/978-1-4612-4272-7

Typeset by the author in TEX

9 8 7 6 5 4 3 2 1

Table of Contents

Chapter 1

Necessary Conditions for Optimality in Problems with Nonstandard Cost Functionals

1. Introduction 1
2. Preliminaries 2
3. Necessary Conditions for Optimality 5
4. Cost Functional of the Form of a Product 10
5. Certain Generalizations 10

References 12

Chapter 2

Synthesis of Suboptimal H_∞ Controllers over a Finite Horizon

Abstract 15

1. Introduction 15
2. Finite Horizon Problem 16
3. Computation of $\tilde{\lambda}$ 22
4. A Differential Equation for $\tilde{\lambda}$ 23
5. Examples 25
6. A Suboptimal Feedback Controller 27
7. Conclusions 30

Appendix 31

References 33

Chapter 3

General Formulae for Suboptimal H_∞ Control over a Finite Horizon

Abstract 35

1. Introduction 35

2. Problem Formulation 36

3. Full State Feedback Problem 37

4. Output Feedback Controller 42

5. Summary of Results 48

6. Conclusions 50

References 50

Chapter 4

Finite Horizon H_∞ with Parameter Variations

Abstract 53

1. Introduction 53

2. Problem Formulation 55

3. Feedback Solutions 56

4. Computation of Performance 58

5. Performance Variation 63

6. Performance Robustness Problem Solution 66

7. An Example 67

8. Conclusions 70

References 71

CHAPTER 5

A General Minimization Problem with Application to Performance Robustness in Finite Horizon H_∞

ABSTRACT 73
1. INTRODUCTION 73
2. EXISTENCE OF A MINIMIZER 74
3. CHARACTERIZATION OF v_0 AND $\hat{\lambda}$ 77
4. VARIATION OF THE MINIMUM VALUE 82
5. APPLICATION TO PERFORMANCE ROBUSTNESS 85
6. CONCLUSIONS 90
REFERENCES 90

CHAPTER 6

H_∞ Design of the F/A-18A Automatic Carrier Landing System

ABSTRACT 93
1. INTRODUCTION 93
2. H_∞ CONTROLLER DESIGN 94
3. ACTUATOR AND ENGINE DYNAMICS 98
4. RESPONSE TO DISTURBANCES 101
5. CONCLUSIONS 104
REFERENCES 104

SUBJECT INDEX 117

Preface

THIS book presents a generalized state-space theory for the analysis and synthesis of finite horizon suboptimal H_∞ controllers. We derive expressions for a suboptimal controller in a general setting and propose an approximate solution to the H_∞ performance robustness problem. The material in the book is taken from a collection of research papers written by the author.

The book is organized as follows. Chapter 1 treats nonlinear optimal control problems in which the cost functional is of the form of a quotient or a product of powers of definite integrals. The problems considered in Chapter 1 are very general, and the results are useful for the computation of the actual performance of an H_∞ suboptimal controller. Such an application is given in Chapters 4 and 5. Chapter 2 gives a criterion for the evaluation of the infimal H_∞ norm in the finite horizon case. Also, a differential equation is derived for the achievable performance as the final time is varied. A general suboptimal control problem is then posed, and an expression for a suboptimal H_∞ state feedback controller is derived. Chapter 3 develops expressions for a suboptimal H_∞ output feedback controller in a very general case via the solution of two dynamic Riccati equations. Assuming the adequacy of linear expressions, Chapter 4 gives an iterative procedure for the synthesis of a suboptimal H_∞ controller that yields the required performance even under parameter variations. As a by-product, an expression for the variation of performance due to parameter variations is given for this specific controller. Chapter 5 treats a general minimization problem in which the cost functional is a quotient of definite integrals. The results are useful in computing the actual performance of a given controller. Also, an expression is given for the variation of performance in terms of variations in the system matrices. In

Chapter 6, a design of the F/A-18A Automatic Carrier Landing System is accomplished using finite horizon H_∞ techniques. Only longitudinal equations of motion are considered, and a suboptimal output feedback controller is synthesized. The object of the design is to maintain a constant flight path angle under worst-case conditions of vertical gust and sensor noise.

I take this opportunity to express my gratitude to the management of the Naval Air Warfare Center for the support of this research. Thanks are also due to my wife Carol and our two year old daughter Susan Rebekah for their love and emotional support. Finally, I would like to give special recognition to our newborn son Matthew Aaron whose arrival fortuitously coincided with the completion of this monograph.

Warminster, Pennsylvania — M. B. S.

January 27, 1995

Finite Horizon H_∞
and Related Control Problems

CHAPTER 1

Necessary Conditions for Optimality in Problems with Nonstandard Cost Functionals

1. INTRODUCTION

Usual formulation of optimal control problems involves the minimization of a cost functional which is of the form of a definite integral. In this chapter we develop necessary conditions for an optimal control in the case of problems in which the cost functional is either a quotient or a product of definite integrals. We call such functionals nonstandard, and these naturally arise in Chapters 2,4, and 5 during the computation of the performance of a suboptimal H_∞ controller. In Chapters 2,4, and 5 a criterion for the evaluation of the cost functional will be presented in the specialized case of linear systems and quadratic integrands. Preliminary results for problems having a fixed final time and free terminal state are in [1]. Related results can also be found in [2,3]. In this monograph, we consider only fixed final time problems. Problems in which the final time is free are treated in [4]. In Section 5, we discuss the relation of our results to those in [2,3].

In this chapter results will be derived for nonlinear systems, although the applications given in Chapters 2,4, and 5 involve only linear systems. We make use of the Dubovitskii-Milyutin formalism [5,6] to derive the necessary conditions. This formalism is narrated in Section 2. We will not give detailed proofs of the results in Section 2 since a very lucid treatment of the theory is given in [6].

We consider problems where the cost functional is of the form of a quotient in Section 3 and of the form of a product in Section 4. Finally, certain generalizations are considered in Section 5.

2. Preliminaries

Throughout this section, unless otherwise stated, E denotes a linear topological space [7]. Let $F(x)$ be a real-valued function defined on E.

DEFINITION 2.1. A vector h is called *a direction of decrease* of $F(x)$ at a point x_0 if we can find a neighborhood U of h, and two numbers $\alpha(F, x_0, h) < 0$ and $\epsilon_0 > 0$ such that

$$F(x_0 + \epsilon h) \leq F(x_0) + \epsilon\alpha \quad \text{for all} \quad 0 < \epsilon < \epsilon_0, \ h \in U.$$

DEFINITION 2.2. A subset $K(\bar{x}) \subset E$ is called *a cone with vertex* $\bar{x}$ if for every $\rho > 0$, $\bar{x} + \rho(x - \bar{x}) \in K(\bar{x})$ whenever $x \in K(\bar{x})$. A cone $K(\bar{x})$ can always be obtained as a translate $\bar{x} + K(0)$ of a cone $K(0)$ with vertex 0. If in addition, $K(\bar{x})$ is convex, then it is called a *convex cone.*

It is easy to verify that the directions of decrease generate an open cone $\tilde{K}(0)$. The functional $F(x)$ is said to be *regularly decreasing* at x_0 if $\tilde{K}(0)$ is convex.

DEFINITION 2.3. The derivative $F'(x_0, h)$ at a point x_0 in the direction of h is given by

$$F'(x_0, h) = \lim_{\epsilon \to 0+} \frac{F(x_0 + \epsilon h) - F(x_0)}{\epsilon}.$$

The following result can be found in [6].

THEOREM 2.1. *Let E be a Banach space and $F(x)$ satisfy a local Lipschitz condition at x_0 (i.e., there exist $\epsilon_0 > 0$ and $\beta > 0$ such that $|F(x_1) - F(x_2)| \leq \beta\|x_1 - x_2\|$ for all $\|x_1 - x_0\| \leq \epsilon_0, \|x_2 - x_0\| \leq \epsilon_0$). Assume that $F(x)$ is differentiable at x_0 in any direction, and $F'(x_0, h)$ is convex as a functional of h (that is, for any $0 \leq \rho \leq 1, F'(x_0, \rho h_1 + (1-\rho)h_2) \leq \rho F'(x_0, h_1) + (1-\rho)F'(x_0, h_2)$). Then $F(x)$ is regularly decreasing at x_0, and $\tilde{K}(0) = \{h \mid F'(x_0, h) < 0\}$.*

Proof. Theorem 7.3 of [6].

DEFINITION 2.4. A nonzero continuous linear functional g is said to be a *support functional* for a set A at $x_0 \in A$ if $g(x) \geq g(x_0)$ for all $x \in A$. Under these conditions, the closed hyperplane $H = \{x \mid g(x) = g(x_0)\}$ is called a *supporting hyperplane* for A at the point x_0.

DEFINITION 2.5. Let $Q \subset E$. A vector h is said to be a *feasible direction* for Q at $x_0 \in E$ if there exists a neighborhood U of h such that $x_0 + \epsilon h \in Q$ for all $h \in U$ and all $0 < \epsilon < \epsilon_0$ for some positive number ϵ_0.

It can be easily verified that the feasible directions generate an open cone K_1 with vertex at 0. We say that Q is *regular in feasible directions* at x_0 if $K_1(0)$ happens to be convex.

The dual space of E (the set of all continuous linear functionals on E) is denoted by E^*. The space E^* is a Banach space with the norm $\|g\| = \sup_{\|x\|\leq 1} |g(x)|, g \in E^*$, if E is a normed linear space. If a cone $K(0) \subset E$, the *dual cone* $K^* = \{g \in E^* \mid g(x) \geq 0 \text{ on } K(0)\}$.

We state the following result on dual cones from [6].

THEOREM 2.2. *Let $Q \in E$ be a closed convex set and $x_0 \in Q$. Let Q^* denote the set of all support functionals for Q at x_0, and K_b the cone of feasible directions for Q at x_0. If* $\text{int}(Q) \neq \emptyset$, *then* $K_b^* = Q^*$.

Proof. Theorem 10.5 of [6].

DEFINITION 2.6. Let $Q \in E$. A vector h is said to be *a tangent direction* to Q at $x_0 \in E$ if we can find $x(\epsilon) \in Q$ for all ϵ between 0 and some $\epsilon_0 > 0$, such that $x(\epsilon) = x_0 + \epsilon h + r(\epsilon)$. The vector $r(\epsilon)$ is such that for any neighborhood U of 0, $(1/\epsilon)r(\epsilon) \in U$ for all sufficiently small $\epsilon > 0$.

It is easily seen that the tangent directions generate a cone with vertex at 0. We say that Q is *regular in tangent directions* at x_0 if the cone of tangent directions to Q at x_0 is convex.

We now give the fundamental theorem due to Dubovitskii and Milyutin [5,6].

THEOREM 2.3. *Let the functional $F(x)$ assume a local minimum on $Q = \bigcup_{i=1}^{n+1} Q_i$ at a point $x_0 \in Q$. Assume that $F(x)$ is regularly decreasing at x_0, with directions of decrease K_0; the inequality constraints $Q_i, i = 1, \ldots, n$ (to be made precise later) are regular in feasible directions at x_0; the equality constraint Q_{n+1} (to be made precise later) is also regular in tangent directions at x_0. Denote the feasible directions for each $Q_i, i = 1, \ldots, n$, by K_i and the tangent directions for Q_{n+1} at x_0 by K_{n+1}. Then there exist $g_i \in K_i^*, i = 0, 1, \ldots, n+1$, not all zero, such that*

$$\sum_{i=0}^{n+1} g_i = 0.$$

Proof. Theorem 6.1 of [6].

To add a note on the notation to be used, $C^n(0,T)$ denotes the space of all n-tuples of real-valued continuous functions on $[0,T]$ with sup-norm topology, and $L_\infty^r(0,T)$ represents the space of all r-tuples of essentially bounded real-valued measurable functions on $[0,T]$ with the usual norm topology. If f is a real-valued function with x as one of its arguments, the partial derivative of f with respect to x is represented by f_x. If B is a real matrix, B^* denotes its transpose. We already used the superscript $*$ to denote dual spaces, dual cones, and so on, and hopefully, no confusion arises on account of this dual usage. The symbol $(\cdot,\cdot)$ represents an ordered pair or inner product, whichever is applicable.

For convenience, we state below a result from [6] which will be subsequently used.

LEMMA 2.1. *Let* $Q = \{x \in L_\infty^r(0,T) \mid x(t) \in M$ *for almost all* $0 \le t \le T, M \subset R^r\}$ *and* $x_0 \in Q$. *If the linear functional defined by*

$$g(x) = \int_0^T (a(t), x(t))\,dt, \qquad a \in L_1^r(0,T),$$

is a support to Q *at the point* x_0, *then* $(a(t), x(t) - x_0(t)) \ge 0$ *for all* $x \in M$ *and almost all* $0 \le t \le T$.

Proof. A simple argument results in the contrapositive. Also, see Example 10.5 of [6].

3. Necessary Conditions for Optimality

In this section we develop necessary conditions for an optimal control for problems in which the final time is fixed. Consider the system

$$\frac{dx}{dt} = f(x(t), u(t), t) \tag{1}$$

with boundary conditions

$$x(0) = c, \tag{2}$$

$$x(T) = d \qquad (T \text{ fixed}), \tag{3}$$

where $x(t) \in R^n, u(t) \in R^r$, and t represent the state vector, the control vector, and time, respectively. The problem is to determine the conditions on $x(t) \in C^n(0,T)$ and $u(t) \in L_\infty^r(0,T)$ that minimize

$$F(x,u) = \frac{\int_0^T \phi^1(x(t), u(t), t)\,dt}{\int_0^T \phi^2(x(t), u(t), t)\,dt} \tag{4}$$

(where ϕ^1 and ϕ^2 are scalar functions), under the constraint

$$u(t) \in M \subset R^r \qquad \text{for almost all} \quad 0 \le t \le T. \tag{5}$$

The case where the domain of definition of the solution is $[t_0, T], t_0 \neq 0$, can be reduced to the above case by a simple substitution of variables.

Let $f(x, u, t)$ and $\phi^i(x, u, t), i = 1, 2$, be continuous in x and u, measurable in t, and continuously differentiable with respect to x and u. Also let $f_u, f_x, \phi_u^i, i = 1, 2, \phi_x^i, i = 1, 2$, be bounded for all bounded (x, u). The set M is assumed to be convex with $\text{int}(M) \neq \emptyset$.

With the above assumptions, let us state the necessary conditions for an optimal control.

THEOREM 3.1. *Let $x^0(t)$ and $u^0(t)$ be optimal and $\int_0^T \phi^2(x^0, u^0, t)\,dt > 0$. Also let $\int_0^T \phi^1(x^0, u^0, t)\,dt$ and $\int_0^T \phi^2(x^0, u^0, t)\,dt$ be finite. Then there exist $\psi(t) \in R^n$ and $\lambda_0 > 0, \lambda_0 \in R^1$, not both identically zero, such that*

$$\frac{d\psi}{dt} = -f_x^*(x^0, u^0, t)\psi(t) + \lambda_0\{\phi_x^1(x^0, u^0, t) - \lambda\phi_x^2(x^0, u^0, t)\}, \tag{6}$$

where

$$\lambda = \frac{\int_0^T \phi^1(x^0, u^0, t)\,dt}{\int_0^T \phi^2(x^0, u^0, t)\,dt}, \tag{7}$$

and, moreover,

$$([-f_u^*(x^0, u^0, t)\psi(t) + \lambda_0\{\phi_u^1(x^0, u^0, t) - \lambda\phi_u^2(x^0, u^0, t)\}], u - u^0(t)) \geq 0 \tag{8}$$

for almost all $0 \leq t \leq T$ and all $u \in M$.

Proof. Let $E = C^n(0, T) \times L_\infty^r(0, T)$. Let Q_2 denote the set of all $(x, u) \in E$ satisfying (1), (2), and (3), and Q_1 the set of all pairs satisfying (5). Regarding Q_1 and Q_2 as inequality and equality constraints, respectively, our problem is to minimize (4) on $Q_1 \cap Q_2$.

A. *Analysis of the functional* $F(x,u)$. By Theorem 2.1, $(\bar{x}(t), \bar{u}(t))$ lies in the cone of directions of decrease if and only if (let $\theta^0(t) = (x^0(t), u^0(t))$, $\gamma^0(t) = (\theta^0(t), t)$)

$$F'(\theta^0, (\bar{x}, \bar{u})) = \frac{\left[\begin{array}{l} [\int_0^T \phi^2(\gamma^0)\,dt] \int_0^T [(\phi_x^1(\gamma^0), \bar{x}) + (\phi_u^1(\gamma^0), \bar{u})]\,dt \\ -[\int_0^T \phi^1(\gamma^0)\,dt] \int_0^T [(\phi_x^2(\gamma^0), \bar{x}) + (\phi_u^2(\gamma^0), \bar{u})]\,dt \end{array}\right]}{[\int_0^T \phi^2(\gamma^0)\,dt]^2} < 0,$$

provided that the denominator is nonzero. Let $\int_0^T \phi^2(x^0, u^0, t)\,dt \neq 0$. Since the denominator is positive, $(\bar{x}, \bar{u})$ lies in K_0 if and only if (simplifying the notation)

$$\left[\int_0^T \phi^2\,dt\right] \int_0^T [(\phi_x^1, \bar{x}) + (\phi_u^1, \bar{u})]\,dt - \left[\int_0^T \phi^1\,dt\right] \int_0^T [(\phi_x^2, \bar{x}) + (\phi_u^2, \bar{u})]\,dt < 0. \tag{9}$$

Let $\int_0^T \phi^1(x^0, u^0, t)\,dt / \int_0^T \phi^2(x^0, u^0, t)\,dt = \lambda$, and without loss of generality, let $\int_0^T \phi^2(x^0, u^0, t)\,dt > 0$. Then (9) can be replaced by

$$\int_0^T [(\phi_x^1, \bar{x}) + (\phi_u^1, \bar{u})]\,dt - \lambda \int_0^T [(\phi_x^2, \bar{x}) + (\phi_u^2, \bar{u})]\,dt < 0.$$

By standard arguments (for example, see [6, Theorem 10.2, p. 69]), if $K_0 \neq \emptyset$, then for any $g_0 \in K_0^*$,

$$g_0(\bar{x}, \bar{u}) = -\lambda_0 \Big\{ \int_0^T [(\phi_x^1, \bar{x}) + (\phi_u^1, \bar{u})]\,dt - \lambda \int_0^T [(\phi_x^2, \bar{x}) + (\phi_u^2, \bar{u})]\,dt \Big\}, \lambda_0 \geq 0. \tag{10}$$

B. *Analysis of the constraint* Q_1. The set Q_1 is closed and convex in E since $Q_1 = C^n(0,T) \times Q_1'$, where $Q_1' = \{u(t) \in L_\infty^r(0,T) \mid u(t) \text{ obeys } (5)\}$ is closed and convex in $L_\infty^r(0,T)$ and has a nonempty interior. Also, $\operatorname{int}(Q_1) \neq \emptyset$. Let K_1 be the cone of feasible directions for Q_1 at (x^0, u^0). Then if $g_1 \in K_1^*$, it follows that (see Theorem 2.2) $g_1 = (0, g_1')$, where $g_1' \in [L_\infty^r(0,T)]^*$ is a support to Q_1' at u^0.

C. *Analysis of the constraint* Q_2. Assume that the nondegeneracy condition $f_u^*(x^0, u^0, t)\psi(t) \neq 0$ holds for any nonzero solution $\psi(t)$ of

$$\frac{d\psi}{dt} = -f_x^*(x^0, u^0, t)\psi(t).$$

Then the tangent subspace K_2 at (x^0, u^0) is the set of all pairs such that

$$\frac{d\bar{x}}{dt} = f_x(x^0, u^0, t)\bar{x} + f_u(x^0, u^0, t)\bar{u}, \qquad \bar{x}(0) = 0, \tag{11}$$

$$\bar{x}(T) = 0. \tag{12}$$

Let $L_1 \subset E, L_2 \subset E$ denote the sets of all $(\bar{x}, \bar{u})$ satisfying (11) and (12), respectively. Then L_1 and L_2 are subspaces, and $K_2 = L_1 \cap L_2$. It is obvious that if $g \in L_2^*$, then $g(\bar{x}, \bar{u}) = (\bar{x}(T), a), a \in R^n$. The space L_2^* is therefore n-dimensional, and $L_1^* + L_2^*$ is weak* closed. Here L_1^* and L_2^* are dual cones. It follows that $K_2^* = L_1^* + L_2^*$. Since L_1 is a subspace, for any $g_2 \in L_1^*, g_2(\bar{x}, \bar{u}) = 0$ for all $(\bar{x}, \bar{u}) \in L_1$. As we already know, if $g_3 \in L_2^*$, then $g_3(\bar{x}, \bar{u}) = (\bar{x}(T), a), a \in R^n$.

D. *Application of Theorem* 2.3. It can be shown that the cone K_1 is convex (see [5,6]). Hence, by Theorem 2.3, there exist $g_0, g_1, g_2, g_3 \in E^*$, not all zero, such that for all $(\bar{x}, \bar{u}) \in E$,

$$g_0 + g_1 + g_2 + g_3 = 0, \tag{13}$$

where g_0 is given by (10), $g_1(\bar{x}, \bar{u}) = g_1'(\bar{u})$ is a support to Q_1' at u^0, $g_2(\bar{x}, \bar{u})$ vanishes for $(\bar{x}, \bar{u})$ satisfying (11), and $g_3(\bar{x}, \bar{u}) = (\bar{x}(T), a), a \in R^n$.

E. *Analysis of Equation* (13). Let $\bar{u}$ be arbitrary, and $\bar{x}(\bar{u})$ be the corresponding solution of (11). Under these conditions $g_2(\bar{x}, \bar{u}) = 0$, and (13) is equivalent to

$$\begin{aligned} g_1'(\bar{u}) = \lambda_0\{ &\int_0^T [(\phi_x^1, \bar{x}) + (\phi_u^1, \bar{u})]\, dt \\ -\lambda &\int_0^T [(\phi_x^2, \bar{x}) + (\phi_u^2, \bar{u})]\, dt\} - (\bar{x}(T), a), \qquad \lambda_0 \geq 0. \end{aligned} \tag{14}$$

Let $\psi(t)$ be the solution of (6) with the boundary condition $\psi(T) = a$. Then it follows that

$$\lambda_0 \int_0^T [(\phi_x^1, \bar{x}) - \lambda(\phi_x^2, \bar{x})]\, dt - (\bar{x}(T), a) = -\int_0^T (f_u^*(x^0, u^0, t)\psi, \bar{u})\, dt.$$

Hence

$$g_1'(\bar{u}) = \int_0^T ([-f_u^*(x^0, u^0, t)\psi + \lambda_0\{\phi_u^1(x^0, u^0, t) - \lambda\phi_u^2(x^0, u^0, t)\}], \bar{u})\, dt,$$

where $\bar{u}$ is arbitrary and $g_1'(\bar{u})$ is a support to Q_1' at u^0. Now, using Lemma 2.1, we have

$$([-f_u^*(x^0, u^0, t)\psi(t) + \lambda_0\{\phi_u^1(x^0, u^0, t) - \lambda\phi_u^2(x^0, u^0, t)\}], u - u^0(t)) \geq 0$$

for almost all $0 \leq t \leq T$ and all $u \in M$, that is, (8) is satisfied.

If $\lambda_0 = 0$ and $\psi(t) \equiv 0$, then we would have $g_i = 0, i = 0, 1, 2, 3$, and this contradicts Theorem 2.3.

F. *Analysis of exceptional cases.* We show that even if $K_0 = \emptyset$ and system (11) is degenerate, the conclusions of Theorem 3.1 are valid. If $K_0 = \emptyset$, then

$$\int_0^T [(\phi_x^1, \bar{x}) + (\phi_u^1, \bar{u})]\, dt - \lambda \int_0^T [(\phi_x^2, \bar{x}) + (\phi_u^2, \bar{u})]\, dt = 0.$$

Choose $\lambda_0 = 1$ and $\psi(T) = 0$. Then for almost all $t \in [0, T]$,

$$-f_u^*(x^0, u^0, t)\psi(t) + \lambda_0\{\phi_u^1(x^0, u^0, t) - \lambda\phi_u^2(x^0, u^0, t)\} = 0,$$

and hence, (8) is satisfied. If (11) is degenerate, choosing $\lambda_0 = 0$ we get a nonzero solution $\psi(t)$ of (6) with $-f_u^*(x^0, u^0, t)\psi(t) \equiv 0$. □

Thus the proof of Theorem 3.1 is complete. If the boundary conditions are such that $x(0) \in S_1, x(T) \in S_2$, where S_1 and S_2 are smooth manifolds in R^n, then the results of Theorem 3.1 are still valid, with the added transversality conditions: $\psi(0)$ and $\psi(T)$ must be orthogonal to the tangent subspaces of S_1 at $x^0(0)$ and S_2 at $x^0(T)$, respectively.

4. Cost Functional of the Form of a Product

Consider the following problem: Find $(x(t), u(t)) \in C^n(0,T) \times L^r_\infty(0,T)$ that minimizes

$$F(x,u) = \left(\int_0^T \phi^1(x(t), u(t), t)\, dt\right)\left(\int_0^T \phi^2(x(t), u(t), t)\, dt\right) \tag{15}$$

under the constraints (1), (2), (3), and (5).

The assumptions for this section are the same as those in Section 3. Following a similar procedure to that in Section 3, the following necessary conditions can be derived.

THEOREM 4.1. *Let $x^0(t)$ and $u^0(t)$ be optimal and $\int_0^T \phi^2(x^0, u^0, t)\, dt > 0$. Also let $\int_0^T \phi^1(x^0, u^0, t)\, dt$ and $\int_0^T \phi^2(x^0, u^0, t)\, dt$ be finite. Then there exist $\psi(t)$ and $\lambda_0 \geq 0$, not both identically zero, such that*

$$\frac{d\psi}{dt} = -f_x^*(x^0, u^0, t)\psi(t) + \lambda_0\{\phi_x^1(x^0, u^0, t) + \lambda\phi_x^2(x^0, u^0, t)\}$$

and

$$([-f_u^*(x^0, u^0, t)\psi(t) + \lambda_0\{\phi_u^1(x^0, u^0, t) + \lambda\phi_u^2(x^0, u^0, t)\}], u - u^0(t)) \geq 0$$

for all $u \in M$ and almost all $0 \leq t \leq T$,

where

$$\lambda = \frac{\int_0^T \phi^1(x^0, u^0, t)\, dt}{\int_0^T \phi^2(x^0, u^0, t)\, dt}.$$

5. Certain Generalizations

Work similar to ours employing variational techniques can be found in [2,3]. The problem treated there involves fixed initial and final times and states. We showed in [4] that similar results can be obtained in the case where the final time is not fixed. Also, our results are applicable to the case involving control constraints, and in general we assume less smoothness on the functions $f(x,u,t)$ and $\phi^i(x,u,t), i = 1,2$.

In [2,3], Miele considers an optimal problem involving products of powers of a finite number of functionals. We will extend our results to this case in the present section. The problem will be the same as the one considered in Section 3 with the cost functional replaced by

$$F(x,u) = \left(\int_0^T \phi^1(x(t),u(t),t)\,dt\right)^{\alpha_1} \left(\int_0^T \phi^2(x(t),u(t),t)\,dt\right)^{\alpha_2}, \qquad (19)$$

where $\alpha_1, \alpha_2 \in R^1$. We impose the same conditions on f, ϕ^1, and ϕ^2 as those in Section 3.

Let $(x^0(t), u^0(t))$ be a solution to the above problem. Assume further that $0 < \int_0^T \phi^i(x^0,u^0,t)\,dt < \infty, i = 1,2$. Then $(x^0(t),u^0(t))$ solves the equivalent problem with the alternate cost functional

$$G(x,u) = \ln F(x,u) = \sum_{i=1}^{2} \alpha_i \ln\left(\int_0^T \phi^i(x(t),u(t),t)\,dt\right). \qquad (20)$$

Note that

$$G'((x^0,u^0),(\bar{x},\bar{u})) = \sum_{i=1}^{2} \frac{\alpha_i \int_0^T [(\phi_x^i,\bar{x}) + (\phi_u^i,\bar{u})]\,dt}{\int_0^T \phi^i(x^0,u^0,t)\,dt},$$

and $G(x,u)$ satisfies the local Lipschitz condition mentioned in Theorem 2.1. Now, mimicking the procedure in Section 3, we get the following lemma.

LEMMA 5.1. *If $(x^0(t),u^0(t))$ is optimal, then there exist $\psi(t)$ and $\lambda_0 \geq 0$, not both identically zero, such that*

$$\frac{d\psi}{dt} = -f_x^*(x^0,u^0,t)\psi + \lambda_0 \sum_{i=1}^{2} \alpha_i \frac{\phi_x^i(x^0,u^0,t)}{\int_0^T \phi^i(x^0,u^0,t)\,dt}$$

and

$$([-f_u^*(x^0,u^0,t)\psi(t)+\lambda_0\{\sum_{i=1}^{2}\frac{\alpha_i\phi_u^i(x^0,u^0,t)}{\int_0^T\phi^i(x^0,u^0,t)\,dt}\}],u-u^0(t))\geq 0$$

for almost all $0\leq t\leq T$ *and all* $u\in M$.

It can be seen that similar results can be obtained if (19) involves a product of powers of more than two (but a finite number) of definite integrals.

We make use of the above results in Chapters 4 and 5 to compute the actual performance of a suboptimal finite horizon H_∞ controller. For now, we present a simple scalar example. We wish to find $x(t)$ that yields the minimum of the cost functional given by $(\int_0^1(\dot{x})^2\,dt)(\int_0^1 x^2\,dt)^{-1}$ under the boundary conditions $x(0)=x(1)=0$. Letting $\dot{x}=u(t)$, the functional to be minimized becomes $(\int_0^1 u^2\,dt)(\int_0^T x^2\,dt)^{-1}$. Applying Theorem 3.1, the solutions that satisfy the boundary conditions are

$$x^0(\lambda,t)=A\sin(\lambda^{1/2}t),\qquad A\neq 0,$$

where $\lambda=n^2\pi^2, n=1,2,\ldots$. The cost for these curves is given by

$$F(x^0,u^0)=\frac{\int_0^1 A^2n^2\pi^2\cos^2(n\pi t)\,dt}{\int_0^1 A^2\sin^2(n\pi t)\,dt}=n^2\pi^2.$$

The minimum value of F is achieved for $n=1$ when $x(t)=A\sin\pi t, A\neq 0$.

References

[1] Subrahmanyam, M. B. and Eyman, E. D., "Optimization with non-standard cost functionals," *Proc. 13th Annual Allerton Conference*, University of Illinois, Urbana, IL, 1975, pp. 168–173.

[2] Miele, A., "The extremization of products of powers of functionals and its application to aerodynamics," *Astronautica Acta*, **12**, No. 1, 1967, pp. 47–51.

[3] MIELE, A., "On the minimization of the product of the powers of several integrals," *Journal of Optimization Theory and Applications*, **1**, No. 2, 1967, pp. 70–82.

[4] SUBRAHMANYAM, M. B., "Necessary conditions for minimum in problems with nonstandard cost functionals," *Journal of Mathematical Analysis and Applications*, **60**, No. 3, 1977, pp. 601–616.

[5] DUBOVITSKII, A. YA. AND MILYUTIN, A. A., "Extremum problems in the presence of restrictions" [English translation], *U.S.S.R. Computational Mathematics and Mathematical Physics*, **5**, No. 3, 1965, pp. 1–80.

[6] GIRSANOV, I. V., *Lectures on Mathematical Theory of Extremum Problems*, Lecture Notes in Economics and Mathematical Systems, No. 67, Springer-Verlag, Berlin, 1972.

[7] KÖTHE, G., *Topological Vector Spaces*, Vol. I, Springer-Verlag, Berlin, 1969.

CHAPTER 2

Synthesis of Suboptimal H_∞ Controllers over a Finite Horizon

ABSTRACT

In this chapter a finite horizon H_∞ optimal control problem is posed and solved. A criterion which is useful for the evaluation of the infimal H_∞ norm in the finite horizon case is given. Also, a differential equation is derived for the measure of performance in terms of the final time. A general suboptimal control problem is then posed, and an expression for a suboptimal controller is derived solving the saddle point conditions. An expression for a feedback controller can be derived by solving a dynamic Riccati equation. Also, a criterion that yields the actual performance of the suboptimal controller is given. In the time-invariant case, the finite horizon controller converges to a static controller as the final time becomes large. Examples are given to illustrate the usefulness of the theory.

1. INTRODUCTION

There are several recent papers attacking the H_∞ problem from a state space point of view [1,3,4,7,13]. There are also finite horizon versions of these problems and extensions have been made to the linear time-varying case as well [7,13]. The state space approach has yielded new insights into the features of the H_∞ controller, and one of these is the separation of the control problem into a full state feedback design and an observer design.

In a different approach taken by this author [9,10], a measure of performance is computed for a given controller, and nonlinear programming algorithms are utilized to find a controller that optimizes the performance.

This approach is suitable for extending the methodology to solve problems involving convex functionals [11]. We have also applied the methodology to solve model reduction problems [12]. One of the main advantages of this approach is the quantification of variation in performance when uncertainties are present in the system matrices. However, it is tedious to compute the optimal controller in this case because it requires several iterations.

In this chapter we consider a finite horizon H_∞ problem. We give a method for the computation of the infimal H_∞ norm in the finite horizon case. This result is very useful since it gives exactly the least upper bound for achievable performance in the finite horizon case. We show by way of examples that the infimal H_∞ norm in the infinite horizon case can be obtained by increasing the final time and computing the infimal H_∞ norm in the finite horizon case. We will also derive a differential equation for the measure of performance as the final time is varied. We then consider a suboptimal H_∞ problem in a generalized case. An expression for a state feedback controller is given in terms of solution of a dynamic Riccati equation. In the time-invariant case, it was observed that the dynamic Riccati equation gave rise to static feedback as the final time was increased. Also, the actual performance of the suboptimal controller is shown to be the least positive value of a parameter for which a certain two point boundary value problem is satisfied. Expressions for a suboptimal output feedback controller will be developed in Chapter 3.

2. Finite Horizon Problem

The linear time-varying system is given by

$$\dot{x} = A(t)x + B_1(t)u + B_2(t)v, \qquad x(t_0) = 0, \tag{1}$$

$$z = C(t)x + D(t)u + E(t)v, \tag{2}$$

where x, u, v, and z represent the state vector, the control vector, the exogenous input vector, and the vector to be controlled, respectively. The matrices $A(t)$, $B_1(t)$, $B_2(t)$, $C(t)$, $D(t)$, and $E(t)$ will be assumed to be continuous on $[t_0, T]$, where T is the final time. In addition, $u, v \in L_2(t_0, T)$. We consider the minimax problem

$$\min_{v \neq 0} \max_{u} \frac{\int_{t_0}^{T} \frac{1}{2} v^*(t) R(t) v(t)\, dt}{\int_{t_0}^{T} \frac{1}{2} z^*(t) W(t) z(t)\, dt}, \tag{3}$$

where $R(t)$ and $W(t)$ are positive definite matrices and the superscript $*$ denotes matrix or vector transpose. The above problem is related to the H_∞ problem since the functional in (3) represents the ratio of the exogenous signal energy to the error energy. We only consider the inputs for which the denominator of (3) is nonzero. We assume that the value of (3) is strictly positive.

Let

$$J(u, v) = \frac{\int_{t_0}^{T} \frac{1}{2} v^*(t) R(t) v(t)\, dt}{\int_{t_0}^{T} \frac{1}{2} z^*(t) W(t) z(t)\, dt}. \tag{4}$$

Using (2), we can write (4) as

$$J(u, v) = \frac{\int_{t_0}^{T} \frac{1}{2} v^*(t) R(t) v(t)\, dt}{\left[\begin{array}{l} \int_{t_0}^{T} \{ \frac{1}{2} x^* W_1 x + x^* W_2 u + \frac{1}{2} u^* W_3 u \\ \qquad\qquad + x^* W_4 v + \frac{1}{2} v^* W_5 v + u^* W_6 v \}\, dt \end{array} \right]}. \tag{5}$$

Notice that the weighting matrices W_1, W_2, W_3, W_4, W_5, and W_6 are time-varying.

Consider the performance criterion

$$\int_{t_0}^{T} \frac{1}{2} v^* R v\, dt - \tilde{\lambda} \int_{t_0}^{T} \frac{1}{2} z^* W z\, dt, \tag{6}$$

where $\tilde{\lambda}$ is the value of (3). We will first find a saddle point (u^0, v^0) with respect to the criterion (6). The functional (6) can be written as

$$\tilde{J}(u,v) = \int_{t_0}^{T} \frac{1}{2} v^* R v \, dt - \tilde{\lambda} \int_{t_0}^{T} \{ \frac{1}{2} x^* W_1 x + x^* W_2 u$$
$$+ \frac{1}{2} u^* W_3 u + x^* W_4 v + \frac{1}{2} v^* W_5 v + u^* W_6 v \} \, dt. \quad (7)$$

Given $u(t)$, let v^0 maximize (7). The following lemma characterizes $v^0(t)$.

LEMMA 2.1. *Let $\tilde{\lambda}$ be such that $R - \tilde{\lambda} W_5$ is positive definite for all $t \in [t_0, T]$. For a given u, if $v^0(t)$ minimizes (7), then there exists an $\eta(t)$ such that*

$$\frac{d\eta}{dt} = -A^* \eta - \tilde{\lambda} W_1 x - \tilde{\lambda} W_2 u - \tilde{\lambda} W_4 v^0, \quad \eta(T) = 0, \quad (8)$$

and

$$v^0(t) = (R - \tilde{\lambda} W_5)^{-1} \{ B_2^* \eta + \tilde{\lambda} W_4^* x + \tilde{\lambda} W_6^* u. \} \quad (9)$$

Proof. By the maximum principle [6], there exists an adjoint response $\eta(t)$ such that the Hamiltonian

$$H = \tilde{\lambda} \{ \frac{1}{2} x^* W_1 x + x^* W_2 u + \frac{1}{2} u^* W_3 u + x^* W_4 v + \frac{1}{2} v^* W_5 v + u^* W_6 v \}$$
$$- \frac{1}{2} v^* R v + \eta^* \{ A(t) x + B_1(t) u + B_2(t) v \} \quad (10)$$

is maximized almost everywhere on $[t_0, T]$. Satisfaction of $\frac{\partial H}{\partial v} = 0$ yields (9). The adjoint variable η satisfies

$$\frac{d\eta}{dt} = -\frac{\partial H}{\partial x} = -A^* \eta - \tilde{\lambda} W_1 x - \tilde{\lambda} W_2 u - \tilde{\lambda} W_4 v^0. \quad (11)$$

By the transversality condition, $\eta(T) = 0$. □

In a similar manner, we can get an expression for an optimal $u^0(t)$ that maximizes (7) for given v.

LEMMA 2.2. *Consider the system given by (1). Assume that W_3 is positive definite for all $t \in [t_0, T]$. For a given v, if $u^0(t)$ maximizes (7), then there exists a $\psi(t)$ such that*

$$\frac{d\psi}{dt} = -A^*\psi + W_1 x + W_2 u^0 + W_4 v, \quad \psi(T) = 0, \tag{12}$$

and

$$u^0(t) = W_3^{-1}\{B_1^*\psi - W_2^* x - W_6 v\}. \tag{13}$$

Proof. Proof is similar to that of Lemma 2.1. □

Simultaneous solution of (8), (9), (12), and (13) yields a saddle point solution (u^0, v^0) for the functional given by (7).

We now express the above minimax solution in a simpler form.

From (8) and (12), at a saddle point solution (u^0, v^0), we get

$$\frac{d}{dt}(\tilde{\lambda}\psi + \eta) = -A^*(t)(\tilde{\lambda}\psi + \eta), \quad \tilde{\lambda}\psi(T) + \eta(T) = 0. \tag{14}$$

It follows that

$$\tilde{\lambda}\psi(t) + \eta(t) = 0, \quad t \in [t_0, T]. \tag{15}$$

Thus the saddle point solution is characterized by

$$\frac{d\psi}{dt} = -A^*\psi + W_1 x + W_2 u^0 + W_4 v^0, \quad \psi(T) = 0, \tag{16}$$

$$u^0(t) = W_3^{-1}\{B_1^*\psi - W_2^* x - W_6 v^0\}, \tag{17}$$

$$v^0(t) = \tilde{\lambda}(R - \tilde{\lambda}W_5)^{-1}\{-B_2^*\psi + W_4^* x + W_6^* u^0\}. \tag{18}$$

We define the optimal controller and the worst disturbance the following way. Note that we do not require W_3 to be positive definite in the following formulae. Assuming that the inverse of $W_3 + \tilde{\lambda}W_6(R - \tilde{\lambda}W_5)^{-1}W_6^*$ exists, let

$$\Omega = (R - \tilde{\lambda} W_5)^{-1}, \tag{19}$$

$$\Lambda = (W_3 + \tilde{\lambda} W_6 \Omega W_6^*)^{-1}, \tag{20}$$

$$U_1 = \Lambda(B_1^* + \tilde{\lambda} W_6 \Omega B_2^*), \tag{21}$$

$$U_2 = -\Lambda(W_2^* + \tilde{\lambda} W_6 \Omega W_4^*), \tag{22}$$

$$V_1 = \tilde{\lambda}\Omega(-B_2^* + W_6^* U_1), \tag{23}$$

$$V_2 = \tilde{\lambda}\Omega(W_4^* + W_6^* U_2). \tag{24}$$

Substituting (18) in (17), we get

$$u^0 = U_1 \psi + U_2 x. \tag{25}$$

Substituting (25) in (18), we get

$$v^0 = V_1 \psi + V_2 x. \tag{26}$$

Thus we have a two point boundary value problem given by

$$\begin{pmatrix} \dot{x} \\ \dot{\psi} \end{pmatrix} = \begin{pmatrix} M & N \\ L & -M^* \end{pmatrix} \begin{pmatrix} x \\ \psi \end{pmatrix}, \tag{27}$$

where

$$M = A + B_1 U_2 + B_2 V_2 = (A^* - W_2 U_1 - W_4 V_1)^*, \tag{28}$$

$$N = B_1 U_1 + B_2 V_1, \tag{29}$$

$$L = W_1 + W_2 U_2 + W_4 V_2, \tag{30}$$

with

$$x(t_0) = 0, \quad \psi(T) = 0. \tag{31}$$

We now give a criterion for the estimation of the parameter $\tilde{\lambda}$. Notice that $\tilde{\lambda} = \min_{v \neq 0} \max_u J(u, v)$ and gives a measure of performance of the optimal controller under worst-case conditions corresponding to $v_0(t)$. In the H_∞ case, the evaluation of $\tilde{\lambda}$ would entail the γ-iteration.

THEOREM 2.1. *Let $\tilde{\lambda}$ be the smallest positive value for which the boundary value problem given by (27) and (31) has a solution (x, ψ) with $\int_{t_0}^{T} \frac{1}{2} z^* W z \, dt > 0$, $z = Cx + Du + Ev$, where $u \stackrel{\text{def}}{=} U_1\psi + U_2 x$ and $v \stackrel{\text{def}}{=} V_1\psi + V_2 x$. Then $\tilde{\lambda}$ is the minimax value of (3), (x, ψ) is an optimal pair, u is an optimal controller, and v is the worst-case exogenous input.*

Proof. It is clear from Lemmas 2.1 and 2.2 that if (u, v) is an optimal pair, then (27) and (31) are satisfied, with $\tilde{\lambda}$ being the minimax value of (3). Now suppose $(x, \psi) \neq 0$ satisfies (27) and (31) for some $\tilde{\lambda}$. Let $u = U_1\psi + U_2 x$ and $v = V_1\psi + V_2 x$. In the following equations, $(\cdot\,,\cdot)$ denotes an inner product.

We have

$$\int_{t_0}^{T} \big((R - \tilde{\lambda} W_5)v, v\big)\, dt = \int_{t_0}^{T} (-\tilde{\lambda} B_2^* \psi, v)\, dt + \tilde{\lambda} \int_{t_0}^{T} \{x^* W_4 v + u^* W_6 v\}\, dt. \tag{32}$$

In the above expression

$$\begin{aligned} \int_{t_0}^{T} (-\tilde{\lambda} B_2^* \psi, v)\, dt &= \int_{t_0}^{T} (-\tilde{\lambda}\psi, B_2 v)\, dt \\ &= -\tilde{\lambda} \int_{t_0}^{T} (\psi, \dot{x} - Ax - B_1 u)\, dt. \end{aligned} \tag{33}$$

An integration by parts and equations (16), (17), and (31) yield

$$-\tilde{\lambda} \int_{t_0}^{T} (B_2^* \psi, v)\, dt = \tilde{\lambda} \int_{t_0}^{T} \{x^* W_1 x + 2x^* W_2 u + u^* W_3 u + x^* W_4 v + u^* W_6 v\}\, dt. \tag{34}$$

Substituting (34) in (32), we get

$$\begin{aligned} \int_{t_0}^{T} v^* R v\, dt = \tilde{\lambda} \int_{t_0}^{T} \{&x^* W_1 x + 2x^* W_2 u + u^* W_3 u \\ &+ 2x^* W_4 v + v^* W_5 v + 2u^* W_6 v\}\, dt. \end{aligned} \tag{35}$$

Thus, the value of (4) associated with (u, v) is $\tilde{\lambda}$ provided that the right side of (35) is nonzero. Hence, if (x, ψ) is a nontrivial solution of the boundary value problem given by (27) and (31) for the smallest parameter $\tilde{\lambda} > 0$ with $\int_{t_0}^{T} z^* W z \, dt > 0$, then $\tilde{\lambda}$ is the value of (3), and (x, ψ) is an optimal pair. □

Note that the boundary value problem given by (27) and (31) has a solution with a nonvanishing denominator for (4) for at most a countably infinite values of $\tilde{\lambda}$. The smallest positive $\tilde{\lambda}$ gives the worst-case performance, and the finite horizon H_∞ norm is precisely $\frac{1}{\sqrt{\tilde{\lambda}}}$. Theorem 2.1 gives a sufficient condition for the pair (u, v) to be optimal.

3. Computation of $\tilde{\lambda}$

In this section we consider the boundary value problem given by (27) and (31) and derive formulas for the computation of $\tilde{\lambda}$.

Making use of the transition matrix, the solution of (27) can be expressed as

$$\begin{pmatrix} x(t) \\ \psi(t) \end{pmatrix} = \begin{pmatrix} \phi_{11}(t, t_0) & \phi_{12}(t, t_0) \\ \phi_{21}(t, t_0) & \phi_{22}(t, t_0) \end{pmatrix} \begin{pmatrix} x(t_0) \\ \psi(t_0) \end{pmatrix}. \tag{36}$$

Substituting the boundary conditions $x(t_0) = 0, \psi(T) = 0$, in (36), we get

$$\psi(T) = 0 = \phi_{22}(T, t_0)\psi(t_0). \tag{37}$$

Thus, we need the least positive $\tilde{\lambda}$ that makes $\det\big(\phi_{22}(T, t_0)\big) = 0$ and the denominator of (4) positive.

We found the following algorithm to be numerically more stable since numbers of lesser magnitude are involved in the computation of the transition matrices in (38). We have

$$\begin{pmatrix} x(T) \\ \psi(T) \end{pmatrix} = \phi(T, \frac{T+t_0}{2})\phi(\frac{T+t_0}{2}, t_0) \begin{pmatrix} x(t_0) \\ \psi(t_0) \end{pmatrix}. \tag{38}$$

In (38), the point $\frac{T+t_0}{2}$ of the interval $[t_0, T]$ is selected since it is the midpoint. Let

$$\phi^{-1}(T, \frac{T+t_0}{2}) = \begin{pmatrix} \xi_{11} & \xi_{12} \\ \xi_{21} & \xi_{22} \end{pmatrix} \tag{39}$$

and

$$\phi(\frac{T+t_0}{2}, t_0) = \begin{pmatrix} \nu_{11} & \nu_{12} \\ \nu_{21} & \nu_{22} \end{pmatrix}. \tag{40}$$

Making use of $x(t_0) = 0$ and $\psi(T) = 0$, we have

$$\begin{pmatrix} \xi_{11} \\ \xi_{21} \end{pmatrix} x(T) = \begin{pmatrix} \nu_{12} \\ \nu_{22} \end{pmatrix} \psi(t_0). \tag{41}$$

The above equation has a nontrivial solution if and only if

$$\det \begin{pmatrix} \xi_{11} & \nu_{12} \\ \xi_{21} & \nu_{22} \end{pmatrix} = 0. \tag{42}$$

Thus, we need the least positive $\tilde{\lambda}$ which makes the above determinant zero.

4. A Differential Equation for $\tilde{\lambda}$

For simplicity we derive a differential equation for $\tilde{\lambda}$ only in the case where $W_4 = W_5 = W_6 = 0$. Thus (27) can be written as

$$\begin{aligned} \dot{x} &= Mx + (B_1 W_3^{-1} B_1^* - \tilde{\lambda} B_2 R^{-1} B_2^*)\psi, \\ \dot{\psi} &= Lx - M^*\psi, \end{aligned} \tag{43}$$

with

$$x(t_0) = 0, \quad \psi(T) = 0. \tag{44}$$

Now assume that the final time is changed to $T + \Delta T$ where ΔT is an elemental increment. The solution of the above boundary value problem can be extended to $[t_0, T + \Delta T]$. Suppose x_1 and ψ_1 are the elemental variations in (x, ψ) owing to the increment ΔT in T. That is, $(x + x_1, \psi + \psi_1)$ is the new optimal pair. Also denote the variation in $\tilde{\lambda}$ by $\Delta\tilde{\lambda}$. We have

$$\begin{aligned} \dot{x}_1 &= Mx_1 + (B_1 W_3^{-1} B_1^* - \tilde{\lambda} B_2 R^{-1} B_2^*)\psi_1 - \Delta\tilde{\lambda} B_2 R^{-1} B_2^* \psi, \\ \dot{\psi}_1 &= Lx_1 - M^*\psi_1, \end{aligned} \tag{45}$$

with

$$x_1(t_0) = 0, \quad \psi_1(T + \Delta T) = -\psi(T + \Delta T). \tag{46}$$

THEOREM 4.1. *As a function of* T, $\tilde{\lambda}$ *satisfies*

$$\frac{d\tilde{\lambda}}{dT} = \frac{-x^*(T)L(T)x(T)}{\int_{t_0}^{T} \psi^* B_2 R^{-1} B_2^* \psi \, dt}. \tag{47}$$

Proof. From (45),

$$\int_{t_0}^{T+\Delta T} x^* \dot{\psi}_1 \, dt = -\int_{t_0}^{T+\Delta T} \{\tilde{\lambda} x^* L x_1 - x^* M^* \psi_1\} \, dt. \tag{48}$$

By an integration by parts,

$$\int_{t_0}^{T+\Delta T} x^* \dot{\psi}_1 \, dt = x^* \psi_1(T + \Delta T) - \int_{t_0}^{T+\Delta T} \{x^* M^* \psi_1 + \psi^* (B_1 W_3^{-1} B_1^* - \tilde{\lambda} B_2 R^{-1} B_2^*) \psi_1\} \, dt. \tag{49}$$

From (48) and (49),

$$\int_{t_0}^{T+\Delta T} x^* L x_1 \, dt = x^* \psi_1(T + \Delta T) - \int_{t_0}^{T+\Delta T} \psi^* (B_1 W_3^{-1} B_1^* - \tilde{\lambda} B_2 R^{-1} B_2^*) \psi_1 \, dt. \tag{50}$$

From (43), the left side of (50) can be written as

$$\int_{t_0}^{T+\Delta T} x^* L x_1 \, dt = \int_{t_0}^{T+\Delta T} (\dot{\psi} + M^* \psi)^* x_1 \, dt. \tag{51}$$

Integrating the right side of (51) by parts, we get

$$\int_{t_0}^{T+\Delta T} x^* L x_1 \, dt = \psi^* x_1(T + \Delta T) - \int_{t_0}^{T+\Delta T} \psi^* (B_1 W_3^{-1} B_1^* - \tilde{\lambda} B_2 R^{-1} B_2^*) \psi_1 \, dt + \Delta\tilde{\lambda} \int_{t_0}^{T+\Delta T} \psi^* B_2 R^{-1} B_2^* \psi \, dt. \tag{52}$$

Substituting (52) in (50), we get

$$\Delta\tilde{\lambda}\int_{t_0}^{T+\Delta T}\psi^* B_2 R^{-1} B_2^* \psi\, dt = x^*(T+\Delta T)\psi_1(T+\Delta T) - \psi^*(T+\Delta T)x_1(T+\Delta T). \tag{53}$$

We have

$$\begin{aligned} x^*\psi_1(T+\Delta T) - \psi^* x_1(T+\Delta T) &= x^*(T)\psi_1(T+\Delta T) - \psi^*(T)x_1(T+\Delta T) + o(\Delta T) \\ &= -x^*(T)\psi(T+\Delta T) + \psi^*(T)x_1(T+\Delta T) + o(\Delta T) \\ &= -x^*(T)\dot{\psi}(T)\Delta T + o(\Delta T) \\ &= -x^*(T)L(T)x(T)\Delta T + o(\Delta T). \end{aligned} \tag{54}$$

From (53) and (54), we get (47). □

5. Examples

The above theory is useful in the computation of the infimal H_∞ norm. This problem is still being researched actively in the case of both static and dynamic controllers, and there are a variety of algorithms in the literature. Let $t_0 = 0$. In the examples given below, we compute the minimum H_∞ norm given by (3) as the final time T varies. The programs were written using PC-MATLAB, and the least positive $\tilde{\lambda}$ which satisfies equation (42) was found. The infimal finite horizon H_∞ norm γ is $\frac{1}{\sqrt{\tilde{\lambda}}}$.

EXAMPLE 1. We consider the tracking example from [2]. In this case

$$A = \begin{pmatrix} -0.1 & 0 & 0 \\ 0 & 0 & 0 \\ 0 & 0 & 2 \end{pmatrix}, \quad B_1 = \begin{pmatrix} 0 \\ 1 \\ 1 \end{pmatrix}, \quad B_2 = \begin{pmatrix} 1 \\ 0 \\ 0 \end{pmatrix},$$

$$W_1 = \begin{pmatrix} 0.0081 & -0.045 & -0.045 \\ -0.045 & 0.25 & 0.25 \\ -0.045 & 0.25 & 0.25 \end{pmatrix}, \quad W_2 = \begin{pmatrix} 0 \\ 0 \\ 0 \end{pmatrix}, \quad W_3 = 1,$$

$$W_4 = \begin{pmatrix} 0.009 \\ -0.05 \\ -0.05 \end{pmatrix}, \quad W_5 = 0.01, \quad W_6 = 0, \quad R = 1.$$

The results are given below. It can be easily verified that $\tilde{\lambda}$ as a function of T obeys Theorem 4.1.

TABLE 1: Results of Example 1		
T	$\tilde{\lambda}$	$\gamma = \frac{1}{\sqrt{\tilde{\lambda}}}$
5	17.6023	0.2384
10	16.3113	0.2476
15	15.9802	0.2502
20	15.7944	0.2516
25	15.7442	0.2520
30	15.7224	0.2522

EXAMPLE 2. This example, taken from [5] has

$$A = \begin{pmatrix} -2 & 1 & 1 & 1 \\ 3 & 0 & 0 & 2 \\ -1 & 0 & -2 & -3 \\ -2 & -1 & 2 & -1 \end{pmatrix}, \quad B_1 = \begin{pmatrix} 0 & 0 \\ 1 & 0 \\ 0 & 0 \\ 0 & 1 \end{pmatrix}, \quad B_2 = \begin{pmatrix} 1 \\ 0 \\ 1 \\ 0 \end{pmatrix}$$

with

$$R = 1, \quad W_1 = \begin{pmatrix} 1 & 0 & -1 & 0 \\ 0 & 0 & 0 & 0 \\ -1 & 0 & 1 & 0 \\ 0 & 0 & 0 & 0 \end{pmatrix}, \quad W_3 = \begin{pmatrix} 1 & 0 \\ 0 & 1 \end{pmatrix},$$

and zero entries in W_2, W_4, W_5, and W_6. Table 2 gives the numerical results.

TABLE 2: Results of Example 2		
T	$\tilde{\lambda}$	$\gamma = \frac{1}{\sqrt{\tilde{\lambda}}}$
5	3.8975	0.5065
10	0.9220	1.0414
15	0.7001	1.1951
20	0.6722	1.2197
25	0.6681	1.2234

EXAMPLE 3. The last example is also taken from [5]. In this case

$$A = \begin{pmatrix} 0 & 1 & 4 & -4 & 1 \\ -3 & -1 & 1 & 2 & 1 \\ 0 & 1 & -1 & -1 & 0 \\ 2 & 1 & -1 & 0 & 1 \\ -1 & 2 & 1 & -2 & -2 \end{pmatrix}, \quad B_1 = \begin{pmatrix} 0 & 0 \\ 4 & 0 \\ 0 & 0 \\ 0 & 0 \\ 0 & 2 \end{pmatrix}, \quad B_2 = \begin{pmatrix} 1 \\ 0 \\ 1 \\ 0 \\ 1 \end{pmatrix}$$

with

$$R = 1, \quad W_1 = \begin{pmatrix} 1 & 0 & 0 & 0 & 0 \\ 0 & 0 & 0 & 0 & 0 \\ 0 & 0 & 1 & 0 & 0 \\ 0 & 0 & 0 & 1 & 0 \\ 0 & 0 & 0 & 0 & 0 \end{pmatrix}, \quad W_3 = \begin{pmatrix} 1 & 0 \\ 0 & 1 \end{pmatrix},$$

with the rest of the matrices having zero entries. The results are given in Table 3.

TABLE 3: Results of Example 3

T	$\tilde{\lambda}$	$\gamma = \frac{1}{\sqrt{\tilde{\lambda}}}$
5	0.89262204	1.05843988
10	0.87538735	1.06880842
15	0.87477493	1.06918248
18	0.87474966	1.06919793

6. A Suboptimal Feedback Controller

The n-dimensional linear time-varying system is given by

$$\dot{x} = A(t)x + B_1(t)u + B_2(t)v, \qquad x(t_0) = 0, \tag{55}$$

$$z = C(t)x + D(t)u + E(t)v. \tag{56}$$

Given $\lambda < \tilde{\lambda}$, consider the functional

$$\int_{t_0}^{T} \frac{1}{2} v^*(t)R(t)v(t)\,dt - \lambda \int_{t_0}^{T} \frac{1}{2} z^*(t)W(t)z(t)\,dt. \tag{57}$$

Following the same notation as in (19)–(24), let

$$\Omega = (R - \lambda W_5)^{-1}, \tag{58}$$

$$\Lambda = (W_3 + \lambda W_6 \Omega W_6^*)^{-1}, \tag{59}$$

$$U_1 = \Lambda(B_1^* + \lambda W_6 \Omega B_2^*), \tag{60}$$

$$U_2 = -\Lambda(W_2^* + \lambda W_6 \Omega W_4^*), \tag{61}$$

$$V_1 = \lambda\Omega(-B_2^* + W_6^* U_1), \tag{62}$$

and

$$V_2 = \lambda\Omega(W_4^* + W_6^* U_2). \tag{63}$$

Define a state feedback controller and an exogenous input by

$$\begin{aligned} u_0 &= U_1 P x + U_2 x, \\ v_0 &= V_1 P x + V_2 x, \end{aligned} \tag{64}$$

where $P(t)$ satisfies the dynamic Riccati equation

$$\begin{aligned} \dot{P} + P(A + B_1 U_2 + B_2 V_2) + (A^* - W_2 U_1 - W_4 V_1)P \\ + P(B_1 U_1 + B_2 V_1)P - (W_1 + W_2 U_2 + W_4 V_2) = 0, \quad P(T) = 0. \end{aligned} \tag{65}$$

The above equation is solved on $[t_0, T]$. The controller feedback gain is then taken to be $U_1 P + U_2$ and the gain of the exogenous input to be $V_1 P + V_2$.

In the Appendix we show that the performance of this feedback controller is greater than λ. We can also get an idea of the actual performance of the state feedback controller given above. For this, let

$$K(t) = U_1(t)P(t) + U_2(t), \tag{66}$$

$$\tilde{W}_1 = W_1 + W_2 K + K^* W_2^* + K^* W_3 K, \tag{67}$$

$$\tilde{W}_2 = W_4 + K^* W_6, \tag{68}$$

and

$$\tilde{W}_3 = W_5. \tag{69}$$

Then the system given by (1) can be written as

$$\dot{x} = (A + B_1 K)x + B_2 v, \quad x(t_0) = 0, \tag{70}$$

and (3) becomes

$$\frac{\int_{t_0}^{T} \frac{1}{2} v^*(t) R(t) v(t)\, dt}{\int_{t_0}^{T} \{\frac{1}{2} x^*(t) \tilde{W}_1(t) x(t) + x^*(t) \tilde{W}_2(t) v(t) + \frac{1}{2} v^*(t) \tilde{W}_3(t) v(t)\}\, dt}. \tag{71}$$

The exogenous input v needs to be selected such that (71) is minimized. Let $\hat{\lambda}$ be the minimum value of (71) over v. A criterion for the evaluation of $\hat{\lambda}$ is given in [9], and we give the main points here. Let

$$\begin{aligned} \hat{A} &= A + B_1 K + \hat{\lambda} B_2 (R - \hat{\lambda} \tilde{W}_3)^{-1} \tilde{W}_2^*, \\ \hat{B} &= B_2 (R - \hat{\lambda} \tilde{W}_3)^{-1} B_2^*, \\ \hat{C} &= -\hat{\lambda} \tilde{W}_1 - \hat{\lambda}^2 \tilde{W}_2 (R - \hat{\lambda} \tilde{W}_3)^{-1} \tilde{W}_2^*. \end{aligned} \tag{72}$$

THEOREM 6.1. *Consider system (70) and assume that there exists an exogenous input which minimizes (71). Now consider the boundary value problem given by*

$$\begin{pmatrix} \dot{x} \\ \dot{\beta} \end{pmatrix} = \begin{pmatrix} \hat{A} & \hat{B} \\ \hat{C} & -\hat{A}^* \end{pmatrix} \begin{pmatrix} x \\ \beta \end{pmatrix}, \tag{73}$$

$$\begin{pmatrix} x(t_0) \\ \beta(T) \end{pmatrix} = \begin{pmatrix} 0 \\ 0 \end{pmatrix}, \tag{74}$$

where $\hat{A}, \hat{B}$, and $\hat{C}$ are as defined by (72). Note that $\hat{\lambda}$ is a parameter in $\hat{A}, \hat{B}$, and $\hat{C}$. Let $R - \hat{\lambda} \tilde{W}_3$ be nonsingular over $[t_0, T]$, and let $v \stackrel{\text{def}}{=} (R - \hat{\lambda} \tilde{W}_3)^{-1} \{B_2^* \beta + \hat{\lambda} \tilde{W}_2^* x\}$. If $\hat{\lambda}$ is the least positive number such that (73)–(74) has a solution $(x(t), \beta(t))$ with $\int_{t_0}^{T} \{\frac{1}{2} x^* \tilde{W}_1 x + x^* \tilde{W}_2 v + \frac{1}{2} v^* \tilde{W}_3 v\}\, dt > 0$, *then $\hat{\lambda}$ is the optimal value. Moreover, x is an optimal trajectory and $v = (R - \hat{\lambda}\tilde{W}_3)^{-1}\{B_2^*\beta + \hat{\lambda}\tilde{W}_2^* x\}$ is an optimal exogenous input.*

Proof. See [9].

We solved equation (65) for Example 2 of Section 5, taking $\lambda = 1/1.3^2$ and $[t_0, T] = [0, 50]$. The eigenvalues of the system matrix A are $-1.3160 \pm 2.9194i, 0.1906$, and -2.5585. The controller feedback gain converged to a constant matrix near $t = 37.5$ and stayed relatively the same until $t = 0$. It is given by

$$\begin{pmatrix} -1.8181 & -2.8017 & -1.9164 & -1.2430 \\ -0.9150 & -1.2430 & -0.8252 & -0.7596 \end{pmatrix}.$$

The eigenvalues corresponding to the state feedback are $-1.4457 \pm 2.9440i$, -2.5903, and -3.1053. These values match well with the results of [5]. Using the criterion of Theorem 6.1, the actual performance $\hat{\lambda}$ of the controller is computed to be 0.6395.

For Example 3 of Section 5, we took $[t_0, T] = [0, 30]$ and λ to be the nearly H_∞-optimal value of $1/1.07^2$. The eigenvalues of A are $-0.0109 \pm 3.7471i$, -3.7138, -1.5906, and 1.3262. The control gain matrix converged near $t = 22.5$ and stayed constant until $t = 0$. It is given by

$$\begin{pmatrix} 13.2076 & -3.6664 & -51.5939 & 61.7644 & 7.9326 \\ -66.5063 & 3.9663 & 189.8827 & -235.8613 & -40.6176 \end{pmatrix}.$$

The eigenvalues of the closed loop system are $-81.5417, -10.5114$, -3.6525, -2.6296, and -1.5654. These are basically the same as the values given in [5]. For this example, $\hat{\lambda} = 0.9183$.

7. Conclusions

In this chapter we presented an open loop solution to the finite horizon H_∞ optimal control problem using saddle point conditions. The optimal value of performance is shown to be the smallest positive value for which a certain boundary value problem possesses a nontrivial solution. Making use of this criterion, the chapter also presents techniques for the computation of the optimal performance, which basically gives the infimal H_∞ norm in the finite horizon case. Then a differential equation for the optimal performance is derived as the final time is varied. Making use of a suboptimal value of performance, a state feedback controller is derived in terms of solution of a dynamic Riccati equation. Also, a criterion for the evaluation of the actual performance of the suboptimal state feedback controller is given. Certain examples which illustrate the theory are worked out.

APPENDIX

In this Appendix we show that for the suboptimal feedback controller given in Section 6,

$$\inf_{v \neq 0} \frac{\int_{t_0}^{T} v^* R v \, dt}{\int_{t_0}^{T} z^* W z \, dt} > \lambda. \tag{A1}$$

An elementary calculation reveals that in equation (65), $B_1 U_1 + B_2 V_1$ and $W_1 + W_2 U_2 + W_4 V_2$ are symmetric, and

$$(A + B_1 U_2 + B_2 V_2)^* = A^* - W_2 U_1 - W_4 V_1. \tag{A2}$$

Thus the solution P of equation (65) is symmetric. Let

$$u_0 = U_1 P x + U_2 x, \tag{A3}$$

$$v_0 = V_1 P x + V_2 x. \tag{A4}$$

Since (25) and (26) follow from (17) and (18), we have

$$u_0 = W_3^{-1} \{ B_1^* P x - W_2^* x - W_6 v_0 \}, \tag{A5}$$

$$v_0 = \lambda (R - \lambda W_5)^{-1} \{ -B_2^* P x + W_4^* x + W_6^* u_0 \}. \tag{A6}$$

Since $P(T) = 0$ and $x(t_0) = 0$, we have

$$\int_{t_0}^{T} \frac{d}{dt}(x^*Px)\,dt = 0. \tag{A7}$$

Since

$$\dot{x} = (A + B_1U_1P + B_1U_2)x + B_2v, \tag{A8}$$

after some algebra we get

$$\begin{aligned}\frac{d}{dt}(x^*Px) &= x^*P\dot{x} + 2x^*P\dot{x} \\ &= x^*W_1x + x^*W_2u_0 + x^*W_4v_0 \\ &\qquad - x^*PB_2v_0 + x^*PB_1u_0 + 2x^*PB_2v. \end{aligned} \tag{A9}$$

From $(A5)$ and $(A6)$, we get

$$B_1^*Px = W_3u_0 + W_2^*x + W_6v_0, \tag{A10}$$

$$B_2^*Px = -\frac{\Omega^{-1}v_0}{\lambda} + W_4^*x + W_6^*u_0. \tag{A11}$$

Substituting $(A10)$ and $(A11)$ in $(A9)$, we get

$$\begin{aligned}\frac{d}{dt}(x^*Px) &= x^*W_1x + 2x^*W_2u_0 + u_0^*W_3u_0 + 2x^*W_4v \\ &\qquad - \frac{v^*\Omega^{-1}v}{\lambda} + 2u_0^*W_6v + \frac{(v-v_0)^*\Omega^{-1}(v-v_0)}{\lambda}. \end{aligned} \tag{A12}$$

From $(A12)$ and $(A7)$, we get

$$\int_{t_0}^{T} v^*Rv\,dt - \lambda\int_{t_0}^{T} z^*Wz\,dt = \int_{t_0}^{T} (v-v_0)^*\Omega^{-1}(v-v_0)\,dt. \tag{A13}$$

Note that in the above equation v is an open loop function of time, whereas v_0 is a feedback function of x. Note also that Ω^{-1} is positive definite on $[0,T]$. Since the map $v \to v_0$ and its inverse are bounded, it follows that there exists $\delta > 0$ such that

$$\int_{t_0}^{T} (v-v_0)\Omega^{-1}(v-v_0)\,dt \geq \delta\int_{t_0}^{T} v^*Rv\,dt. \tag{A14}$$

Equations ($A13$) and ($A14$) yield

$$\inf_{v \neq 0} \frac{\int_{t_0}^{T} v^* R v \, dt}{\int_{t_0}^{T} z^* W z \, dt} > \lambda.$$

References

[1] Doyle, J. C., Glover, K., Khargonekar, P. P., and Francis, B. A., "State-space solutions to standard H_2 and H_∞ control problems," *IEEE Transactions on Automatic Control*, **34**, 1989, pp. 831–847.

[2] Francis, B. A. and Doyle, J. C., "Linear Control Theory with an H_∞ Optimality Criterion," *SIAM J. Control and Optimization*, **25**, 1987, pp. 815–844.

[3] Khargonekar, P. P., "State-space H_∞ Control Theory," in *Mathematical System Theory: The Influence of R. E. Kalman*, edited by A. C. Antoulas, Springer-Verlag, Berlin, 1991.

[4] Khargonekar, P. P., Nagpal, K. M., and Poolla, K. R., "H_∞ control with transients," *SIAM Journal on Control and Optimization*, **29**, 1991, pp. 1373–1393.

[5] Perkins, W. R. and Medanic, J. V., "Systematic Low Order Controller Design for Disturbance Rejection with Plant Uncertainties," *Report #WRDC-TR-90-3036*, Wright-Patterson AFB, U.S.A., 1990.

[6] Pontryagin, L. S., Boltyanski, V. G., Gamkrelidze, R. V., and Mischenko, E. F., *The Mathematical Theory of Optimal Processes*, Interscience Publishers, New York, 1962.

[7] Ravi, R., Nagpal, K. M., and Khargonekar, P. P., "H^∞ control of linear time-varying systems: A state space approach," *SIAM Journal on Control and Optimization*, **29**, 1991, pp. 1394–1413.

[8] SUBRAHMANYAM, M. B., "Optimal disturbance rejection and performance robustness in linear systems," *Journal of Mathematical Analysis and Applications*, **164**, 1992, pp. 130–150.

[9] SUBRAHMANYAM, M. B., *Optimal Control with a Worst-Case Performance Criterion and Applications*, Lecture Notes in Control and Information Sciences, No. 145, Springer-Verlag, Berlin, 1990.

[10] SUBRAHMANYAM, M. B., "Synthesis of finite-interval H_∞ controllers by state-space methods," *AIAA Journal of Guidance, Control, and Dynamics*, **13**, 1990, pp. 624–629.

[11] SUBRAHMANYAM, M. B., "Worst-case performance measures for linear control problems," *Proc. IEEE Conference on Decision and Control*, Honolulu, U.S.A., 1990, pp. 2439–2443.

[12] SUBRAHMANYAM, M. B. AND STEINBERG, M., "Model reduction with a finite-interval H_∞ criterion," *Proc. AIAA Guidance, Navigation and Control Conference*, Portland, U.S.A., 1990, pp. 1419–1427.

[13] TADMOR, G., "Worst-case design in the time domain: the maximum principle and the standard H_∞ problem," *Mathematics of Control, Signals, and Systems*, **3**, 1990, pp. 301–324.

CHAPTER 3

General Formulae for Suboptimal H_∞ Control over a Finite Horizon

ABSTRACT

In this chapter a general suboptimal control problem is posed, and an expression for a suboptimal controller is derived solving the saddle point conditions. Based on this, a formula for a state feedback suboptimal controller can be derived by solving a dynamic Riccati equation. Then an expression for a suboptimal output feedback controller is developed in a general case via the solution of two dynamic Riccati equations.

1. INTRODUCTION

There are several recent papers attacking the H_∞ problem from a state space point of view [1–5]. There are also finite horizon versions of these problems, and extensions have been made to the linear time-varying case as well [4,5]. The state space approach has yielded new insights into the features of the H_∞ controller, and one of these is the separation of the control problem into a full state feedback design and an observer design.

In a different approach taken by this author [6–8], a measure of performance is computed for a given controller, and nonlinear programming algorithms are utilized to find a controller that optimizes the performance. This approach is suitable for solving problems involving convex functionals [9]. We have also applied the methodology to solve model reduction problems [10]. One of the main advantages of this approach is the quantification of variation in performance when uncertainties are present in the system matrices. However, it is tedious to compute the optimal controller in this case because doing so would require several iterations.

Even though the suboptimal H_∞ problem has been solved, expressions for the controller are not available in the general case. Various assumptions need to be made to derive the expressions for the controller. These assumptions can be removed by means of various transformations [11].

The main contribution of this chapter is the derivation of a suboptimal controller in the time-varying case for a general performance index. We consider a generalized finite horizon suboptimal H_∞ problem. An expression for a state feedback controller is given in terms of solution of a dynamic Riccati equation. Also, an expression for a suboptimal output feedback controller is developed in terms of solutions of two dynamic Riccati equations. Throughout the chapter an objective has been to derive results in as general a case as feasible. The formulae for the synthesis of a suboptimal H_∞ controller are summarized in Section 5, and these can be programmed easily on a digital computer to synthesize a suboptimal H_∞ controller. In the time-invariant case, if the final time is sufficiently large, the solutions of the dynamic Riccati equations converge to the solutions of the corresponding algebraic Riccati equations.

2. Problem Formulation

Let the n-dimensional time-varying system be given by

$$\dot{x} = A(t)x + B_1(t)u + B_2(t)v, \quad x(t_0) = 0, \tag{1}$$

$$z = C(t)x + D(t)u + E(t)v, \tag{2}$$

$$y = C_2(t)x + D_2(t)u + E_2(t)v. \tag{3}$$

Without loss of generality, let $t_0 = 0$. Also let

$$\lambda_{\text{opt}} = \max_u \min_{v \neq 0} \frac{\int_0^T \frac{1}{2} v^* R v \, dt}{\int_0^T \frac{1}{2} z^* W z \, dt}, \tag{4}$$

where R and W are assumed to be positive definite and the superscript $*$ denotes a matrix or vector transpose. Computational techniques for the evaluation of λ_{opt} are given in Chapter 2 and [12]. The problems addressed in the chapter can be stated as follows.

PROBLEM 1. Given $\lambda < \lambda_{\text{opt}}$, find a full state feedback controller, if one exists, for which

$$\min_{v \neq 0} \frac{\int_0^T \frac{1}{2} v^* R v \, dt}{\int_0^T \frac{1}{2} z^* W z \, dt} > \lambda.$$

PROBLEM 2. Given $\lambda < \lambda_{\text{opt}}$, find an output feedback controller, if one exists, for which

$$\min_{v \neq 0} \frac{\int_0^T \frac{1}{2} v^* R v \, dt}{\int_0^T \frac{1}{2} z^* W z \, dt} > \lambda.$$

3. Full State Feedback Problem

Consider the performance criterion

$$\int_0^T \frac{1}{2} v^* R v \, dt - \lambda \int_0^T \frac{1}{2} z^* W z \, dt \tag{5}$$

where $0 < \lambda < \infty$. We will first find a saddle point (u^0, v^0), if one exists, with respect to the criterion (5). The motivation for finding the saddle point is that from the expressions of the saddle point, we can construct a suboptimal H_∞ controller on the finite horizon $[0, T]$.

The functional (5) can be written as

$$J(u, v) = \int_0^T \frac{1}{2} v^* R v \, dt - \lambda \int_0^T \{ \frac{1}{2} x^* W_1 x + x^* W_2 u + \frac{1}{2} u^* W_3 u$$
$$+ x^* W_4 v + \frac{1}{2} v^* W_5 v + u^* W_6 v \} \, dt. \tag{6}$$

Given $u(t)$, let v^0 maximize (6). The following lemma characterizes $v^0(t)$.

LEMMA 3.1. *Let λ be such that $R - \lambda W_5$ is positive definite for all $t \in [0, T]$. For a given u, if $v^0(t)$ minimizes (6), then there exists an $\eta(t)$ such that*

$$\frac{d\eta}{dt} = -A^*\eta - \lambda W_1 x - \lambda W_2 u - \lambda W_4 v^0, \quad \eta(T) = 0, \tag{7}$$

and

$$v^0(t) = (R - \lambda W_5)^{-1}\{B_2^*\eta + \lambda W_4^* x + \lambda W_6^* u\}. \tag{8}$$

Proof. By the maximum principle [13], there exists an adjoint response $\eta(t)$ such that the Hamiltonian

$$\begin{aligned} H = \lambda\{\frac{1}{2}x^*W_1x + x^*W_2u + \frac{1}{2}u^*W_3u + x^*W_4v + \frac{1}{2}v^*W_5v + u^*W_6v\} \\ -\frac{1}{2}v^*Rv + \eta^*\{A(t)x + B_1(t)u + B_2(t)v\} \end{aligned} \tag{9}$$

is maximized almost everywhere on $[0, T]$. Satisfaction of $\frac{\partial H}{\partial v} = 0$ yields (8). The adjoint variable η satisfies

$$\frac{d\eta}{dt} = -\frac{\partial H}{\partial x} = -A^*\eta - \lambda W_1 x - \lambda W_2 u - \lambda W_4 v^0. \tag{10}$$

By the transversality condition, $\eta(T) = 0$. □

In a similar manner, we can get an expression for an optimal $u^0(t)$ which maximizes (6) for given v and λ.

LEMMA 3.2. *Consider the system given by (1). Assume that W_3 is positive definite for all $t \in [0, T]$. For a given v, if $u^0(t)$ maximizes (6), then there exists a $\psi(t)$ such that*

$$\frac{d\psi}{dt} = -A^*\psi + W_1 x + W_2 u^0 + W_4 v, \quad \psi(T) = 0, \tag{11}$$

and

$$u^0(t) = W_3^{-1}\{B_1^*\psi - W_2^* x - W_6 v\}. \tag{12}$$

Proof. Proof is similar to that of Lemma 3.1. □

Simultaneous solution of (7), (8), (11), and (12) yields a saddle point solution (u^0, v^0) for the functional given by (6). We now express the above minimax solution in a simpler form.

From (7) and (11), at a saddle point solution (u^0, v^0) we get

$$\frac{d}{dt}(\lambda\psi + \eta) = -A^*(t)(\lambda\psi + \eta), \quad \lambda\psi(T) + \eta(T) = 0. \tag{13}$$

It follows that

$$\lambda\psi(t) + \eta(t) = 0, \quad t \in [0, T]. \tag{14}$$

Thus the saddle point solution is characterized by

$$\frac{d\psi}{dt} = -A^*\psi + W_1 x + W_2 u^0 + W_4 v^0, \quad \psi(T) = 0, \tag{15}$$

$$u^0(t) = W_3^{-1}\{B_1^*\psi - W_2^* x - W_6 v^0\}, \tag{16}$$

$$v^0(t) = \lambda(R - \lambda W_5)^{-1}\{-B_2^*\psi + W_4^* x + W_6^* u^0\}. \tag{17}$$

Let $0 < \lambda < \lambda_{\text{opt}}$. We define a *full state feedback controller* in the following way. Note that we do not require W_3 to be positive definite in the following formulae. Assuming that the inverse of $W_3 + \lambda W_6(R - \lambda W_5)^{-1} W_6^*$ exists, let

$$\Omega = (R - \lambda W_5)^{-1}, \tag{18}$$

$$\Lambda = (W_3 + \lambda W_6 \Omega W_6^*)^{-1}, \tag{19}$$

$$U_1 = \Lambda(B_1^* + \lambda W_6 \Omega B_2^*), \tag{20}$$

$$U_2 = -\Lambda(W_2^* + \lambda W_6 \Omega W_4^*), \tag{21}$$

$$V_1 = \lambda\Omega(-B_2^* + W_6^* U_1), \tag{22}$$

$$V_2 = \lambda\Omega(W_4^* + W_6^* U_2). \tag{23}$$

Substituting (17) in (16), we get

$$u^0 = U_1\psi + U_2x. \tag{24}$$

Substituting (24) in (17), we get

$$v^0 = V_1\psi + V_2x. \tag{25}$$

Let $\epsilon > 0$ be arbitrarily small. On $[\epsilon, T]$, let

$$\psi(t) = P(t)x(t). \tag{26}$$

On $[\epsilon, T]$, we get

$$\begin{aligned}\dot{P} &+ P(A + B_1U_2 + B_2V_2) + (A^* - W_2U_1 - W_4V_1)P \\ &+P(B_1U_1 + B_2V_1)P - (W_1 + W_2U_2 + W_4V_2) = 0, \quad P(T) = 0. \end{aligned}\tag{27}$$

Define the feedback controller and the exogenous input by

$$u_0 = U_1Px + U_2x, \tag{28}$$

$$v_0 = V_1Px + V_2x, \tag{29}$$

where P is the solution of (27) on $[0, T]$. We now show that the performance of this feedback controller is greater than λ.

THEOREM 3.1. *Consider equations (27) and (28). Then for this controller*

$$\inf_{v\neq 0} \frac{\int_0^T \frac{1}{2}v^*Rv\,dt}{\int_0^T \frac{1}{2}z^*W z\,dt} > \lambda. \tag{30}$$

Proof. An elementary calculation shows that in equation (27), both $B_1U_1 + B_2V_1$ and $W_1 + W_2U_2 + W_4V_2$ are symmetric, and

$$(A + B_1U_2 + B_2V_2)^* = A^* - W_2U_1 - W_4V_1. \tag{31}$$

Thus P is symmetric. Since (24) and (25) follow from (16) and (17), we have

$$u_0 = W_3^{-1}\{B_1^* Px - W_2^* x - W_6 v_0\}, \tag{32}$$

$$v_0 = \lambda\Omega\{-B_2^* Px + W_4^* x + W_6^* u_0\}. \tag{33}$$

Since $P(T) = 0$ and $x(0) = 0$, we have

$$\int_0^T \frac{d}{dt}(x^* Px)\, dt = 0. \tag{34}$$

Since

$$\dot{x} = Ax + B_1 u_0 + B_2 v = (A + B_1 U_1 P + B_1 U_2)x + B_2 v, \tag{35}$$

where v is an arbitrary function of time, after some algebra we get

$$\begin{aligned}\frac{d}{dt}(x^* Px) &= x^* \dot{P} x + 2x^* P\dot{x} \\ &= x^* W_1 x + x^* W_2 u_0 + x^* W_4 v_0 - x^* P B_2 v_0 \\ &\qquad + x^* P B_1 u_0 + 2x^* P B_2 v.\end{aligned} \tag{36}$$

From (32) and (33), we get

$$B_1^* Px = W_3 u_0 + W_2^* x + W_6 v_0, \tag{37}$$

$$B_2^* Px = -\frac{\Omega^{-1} v_0}{\lambda} + W_4^* x + W_6 u_0. \tag{38}$$

Substituting (37) and (38) in (36), we get

$$\begin{aligned}\frac{d}{dt}(x^* Px) &= x^* W_1 x + 2x^* W_2 u_0 + u_0^* W_3 u_0 + 2x^* W_4 v \\ &\quad - \frac{v^* \Omega^{-1} v}{\lambda} + 2u_0^* W_6 v + \frac{(v - v_0)^* \Omega^{-1}(v - v_0)}{\lambda}.\end{aligned} \tag{39}$$

From (39) and (34), we get

$$\int_0^T v^* R v\, dt - \lambda \int_0^T z^* W z\, dt = \int_0^T (v - v_0)^* \Omega^{-1} (v - v_0)\, dt. \tag{40}$$

Note that in the above equation, v is an open loop function of t, whereas v_0 is a feedback function of x. Note also that Ω^{-1} is positive definite on $[0, T]$. Since the map $v \to v - v_0$ and its inverse are bounded, it follows that there exists $\delta > 0$ such that

$$\int_0^T (v - v_0)^* \Omega^{-1} (v - v_0)\, dt \geq \delta \int_0^T v^* R v\, dt. \tag{41}$$

From (40) and (41), we have (30). □

4. Output Feedback Controller

Consider again the system given by

$$\dot{x} = A(t)x + B_1(t)u + B_2(t)v, \tag{42}$$

$$z = C(t)x + D(t)u + E(t)v, \tag{43}$$

$$y = C_2(t)x + D_2(t)u + E_2(t)v. \tag{44}$$

Let

$$\tilde{C} = C_2 + E_2 V_1 P + E_2 V_2, \tag{45}$$

where V_1 and V_2 are defined by (22) and (23). Assume that the controller is of the form

$$\dot{q} = Aq + B_1(U_1 P q + U_2 q) + B_2(V_1 P q + V_2 q) + L(\tilde{C} q + D_2 u - y), \tag{46}$$

$$u = U_1 P q + U_2 q, \tag{47}$$

where the observer gain L needs to be determined.

Let

$$u_0 = U_1 P x + U_2 x, \tag{48}$$

$$v_0 = V_1 P x + V_2 x, \tag{49}$$

$$r = u - u_0, \tag{50}$$

$$w = v - v_0, \tag{51}$$

$$e = x - q. \tag{52}$$

Let the control be given by (47), and let P be the solution of (27). We have

$$\dot{x} = Ax + B_1(U_1 P + U_2)q + B_2 v, \tag{53}$$

where v is an arbitrary function of time.

LEMMA 4.1. *For the above system, we have*

$$\int_0^T v^* R v\, dt - \lambda \int_0^T z^* W z\, dt = \int_0^T w^* \Omega^{-1} w\, dt$$
$$-\lambda \int_0^T r^* W_3 r\, dt - 2\lambda \int_0^T r^* W_6 w\, dt. \tag{54}$$

Proof. We have

$$x^* \dot{P} x + 2x^* P \dot{x} = x^* W_1 x + x^* W_2 u_0 + x^* W_4 v_0$$
$$+x^* P B_1(2u - u_0) + x^* P B_2(2v - v_0). \tag{55}$$

From (37) and (38),

$$B_1^* P x = W_3 u_0 + W_2^* x + W_6 v_0, \tag{56}$$

$$B_2^* P x = -\frac{\Omega^{-1} v_0}{\lambda} + W_4^* x + W_6^* u_0. \tag{57}$$

Incorporating (56) and (57) in (55) and rearranging, the right side of (55) can be written as

$$x^* W_1 x + 2x^* W_2 u + u^* W_3 u + 2x^* W_4 v - \frac{v^* \Omega^{-1} v}{\lambda} + 2u^* W_6 v$$
$$-(u - u_0)^* W_3 (u - u_0) + \frac{(v - v_0)^* \Omega^{-1} (v - v_0)}{\lambda} - 2(u - u_0)^* W_6 (v - v_0). \tag{58}$$

Since

$$\int_0^T \frac{d}{dt}(x^* P x)\, dt = \int_0^T \{x^* \dot{P} x + 2x^* P \dot{x}\}\, dt = 0, \tag{59}$$

the result follows. □

We want to ultimately show that the right side of (54) is larger than $\tilde{\delta}\int_0^T v^*Rv\,dt$ for some $\tilde{\delta} > 0$. It easily follows that $e = x - q$ and $r = u - u_0$ satisfy

$$\dot{e} = (\tilde{A} + L\tilde{C})e + (B_2 + LE_2)w, \tag{60}$$

$$r = \tilde{B}e, \tag{61}$$

where

$$\tilde{A} = A + B_2V_1P + B_2V_2, \tag{62}$$

$$\tilde{B} = -(U_1P + U_2). \tag{63}$$

Note that $\tilde{C}$ is defined by (45) and L is the gain of the observer. The right side of (54) can be written as

$$\int_0^T w^*Rw\,dt - \lambda\int_0^T z_1^*Wz_1\,dt, \tag{64}$$

where (see (2))

$$z_1 = Dr + Ew = D\tilde{B}e + Ew. \tag{65}$$

We determine L by considering the dual of (60) and (65). Our ultimate goal is to show that for the L to be chosen, the controller given by (46) and (47) is suboptimal.

Let $\tau = -t$. The dual system can be defined on $[-T, 0]$ as

$$\frac{d\tilde{e}}{d\tau} = \tilde{A}^*\tilde{e} + \tilde{C}^*\tilde{u} + \tilde{B}^*D^*\tilde{w}, \tag{66a}$$

$$\tilde{z}_1 = B_2^*\tilde{e} + E_2^*\tilde{u} + E^*\tilde{w}, \tag{66b}$$

and the functional corresponding to (64) is written as

$$\int_{-T}^0 \frac{1}{2}\tilde{w}^*W^{-1}\tilde{w}\,d\tau - \lambda\int_{-T}^0 \frac{1}{2}\tilde{z}_1^*R^{-1}\tilde{z}_1\,d\tau. \tag{67}$$

Let us write (67) as

$$\int_{-T}^{0} \frac{1}{2}\tilde{w}^* W^{-1}\tilde{w}\, d\tau - \lambda \int_{-T}^{0} \{\frac{1}{2}\tilde{e}^*\tilde{W}_1\tilde{e} + \tilde{e}^*\tilde{W}_2\tilde{u} + \frac{1}{2}\tilde{u}^*\tilde{W}_3\tilde{u}$$
$$+\tilde{e}^*\tilde{W}_4\tilde{w} + \frac{1}{2}\tilde{w}^*\tilde{W}_5\tilde{w} + \tilde{u}^*\tilde{W}_6\tilde{w}\}\, d\tau. \tag{68}$$

We now find a saddle point solution for the functional in (68), lifting for the moment the restrictions on λ. The saddle point solution will be useful in defining the gain L of the observer.

This problem is completely analogous to the full state feedback problem of Section 3, and we can write the solution by inspection. Equations (66a) and (68) are analogous to equations (1) and (6), respectively. Assume that $\tilde{W}_3$ and $W^{-1} - \lambda\tilde{W}_5$ are positive definite and β is the adjoint variable. Observing (15)–(17), the saddle point solution is characterized by

$$\frac{d\beta}{d\tau} = -\tilde{A}\beta + \tilde{W}_1\tilde{e} + \tilde{W}_2\tilde{u} + \tilde{W}_4\tilde{w}, \quad \beta(0) = 0, \tag{69}$$
$$\tilde{u} = \tilde{W}_3^{-1}\{\tilde{C}\beta - \tilde{W}_2^*\tilde{e} - \tilde{W}_6^*\tilde{w}\}, \tag{70}$$
$$\tilde{w} = \lambda(W^{-1} - \lambda\tilde{W}_5)^{-1}\{-D\tilde{B}\beta + \tilde{W}_4^*\tilde{e} + \tilde{W}_6^*\tilde{u}\}. \tag{71}$$

The assumption that $\tilde{W}_3$ be positive definite can be relaxed. Assume that $\tilde{W}_3 + \lambda\tilde{W}_6(W^{-1} - \lambda\tilde{W}_5)^{-1}\tilde{W}_6^*$ is invertible. The equations analogous to (18)–(23) are

$$\Phi = (W^{-1} - \lambda\tilde{W}_5)^{-1}, \tag{72}$$
$$\Gamma = (\tilde{W}_3 + \lambda\tilde{W}_6\Phi\tilde{W}_6^*)^{-1}, \tag{73}$$
$$S_1 = \Gamma(\tilde{C} + \lambda\tilde{W}_6\Phi D\tilde{B}), \tag{74}$$
$$S_2 = -\Gamma(\tilde{W}_2^* + \lambda\tilde{W}_6\Phi\tilde{W}_4^*), \tag{75}$$
$$T_1 = \lambda\Phi(-D\tilde{B} + \tilde{W}_6^* S_1), \tag{76}$$
$$T_2 = \lambda\Phi(\tilde{W}_4^* + \tilde{W}_6^* S_2). \tag{77}$$

Substituting (71) in (70), we can write $\tilde{u}$ and $\tilde{w}$ as

$$\tilde{u} = S_1\beta + S_2\tilde{e}, \tag{78}$$

$$\tilde{w} = T_1\beta + T_2\tilde{e}. \tag{79}$$

Letting

$$\beta = Y\tilde{e} \tag{80}$$

on $[-T, 0]$, we get the Riccati equation

$$-\frac{dY}{d\tau} = (\tilde{A} - \tilde{W}_2S_1 - \tilde{W}_4T_1)Y + Y(\tilde{A}^* + \tilde{C}^*S_2 + \tilde{B}^*D^*T_2)$$
$$+Y(\tilde{C}^*S_1 + \tilde{B}^*D^*T_1)Y - (\tilde{W}_1 + \tilde{W}_2S_2 + \tilde{W}_4T_2), \quad Y(0) = 0, \tag{81}$$

which is analogous to (27). It can be easily verified that the above equation is symmetric.

Now let $0 < \lambda < \lambda_{\text{opt}}$. Define $\tilde{u}_0$ and $\tilde{w}_0$ by

$$\tilde{u}_0 = S_1Y\tilde{e} + S_2\tilde{e}, \tag{82}$$

$$\tilde{w}_0 = T_1Y\tilde{e} + T_2\tilde{e}. \tag{83}$$

LEMMA 4.2. *Consider equations (81) and (82), and assume that at $\tilde{u} = \tilde{u}_0$, given $\epsilon > 0$ there exists $\tilde{T} > 0$ such that for any $\tilde{w} \neq 0$ and all $T \geq \tilde{T}$, $|\tilde{e}^*(-T)Y(-T)\tilde{e}(-T)| \leq \epsilon \int_{-T}^{0} \tilde{w}^*W^{-1}\tilde{w}\, d\tau$. Then there exists $\delta > 0$ such that for all $T \geq \tilde{T}$,*

$$\int_{-T}^{0} \tilde{w}^*W^{-1}\tilde{w}\, d\tau - \lambda \int_{-T}^{0} \tilde{z}_1^*R^{-1}\tilde{z}_1\, d\tau \geq \delta \int_{-T}^{0} \tilde{w}^*W^{-1}\tilde{w}\, d\tau. \tag{84}$$

Proof. Following similar reasoning as in the proof of Theorem 3.1, we can write the equation analogous to (39) as

$$\frac{d}{d\tau}(\tilde{e}^*Y\tilde{e}) = \tilde{e}^*\tilde{W}_1\tilde{e} + 2\tilde{e}^*\tilde{W}_2\tilde{u}_0 + \tilde{u}_0^*\tilde{W}_3\tilde{u}_0 + 2\tilde{e}^*\tilde{W}_4\tilde{w} + \tilde{w}^*\tilde{W}_5\tilde{w}$$
$$+2\tilde{u}_0^*\tilde{W}_6\tilde{w} - \frac{\tilde{w}^*W^{-1}\tilde{w}}{\lambda} + \frac{(\tilde{w} - \tilde{w}_0)^*\Phi^{-1}(\tilde{w} - \tilde{w}_0)}{\lambda}. \tag{85}$$

Integrating both sides from $-T$ to 0, we get

$$\int_{-T}^{0} \tilde{w}^* W^{-1} \tilde{w}\, d\tau - \lambda \int_{-T}^{0} \tilde{z}_1^* R^{-1} \tilde{z}_1\, d\tau =$$

$$\int_{-T}^{0} (\tilde{w} - \tilde{w}_0)^* \Phi^{-1} (\tilde{w} - \tilde{w}_0)\, d\tau + \lambda \tilde{e}^*(-T) Y(-T) \tilde{e}(-T). \quad (86)$$

Let $\epsilon > 0$ be sufficiently small. Since the map $\tilde{w} \to \tilde{w} - \tilde{w}_0$ and its inverse are bounded, there exists $\delta > 0$ such that

$$\int_{-T}^{0} (\tilde{w} - \tilde{w}_0)^* \Phi^{-1} (\tilde{w} - \tilde{w}_0)\, d\tau > (\delta + \lambda\epsilon) \int_{-T}^{0} \tilde{w}^* W^{-1} \tilde{w}\, d\tau. \quad (87)$$

Now by the assumption on $\tilde{e}^*(-T)Y(-T)\tilde{e}(-T)$, (84) follows, if $T \geq \tilde{T}$. □

We now go back to the original problem and reverse time in (81).

THEOREM 4.1. *On* $[0, T]$, *let*

$$\dot{Y} = (\tilde{A} - \tilde{W}_2 S_1 - \tilde{W}_4 T_1) Y + Y(\tilde{A}^* + \tilde{C}^* S_2 + \tilde{B}^* D^* T_2)$$

$$+ Y(\tilde{C}^* S_1 + \tilde{B}^* D^* T_1) Y - (\tilde{W}_1 + \tilde{W}_2 S_2 + \tilde{W}_4 T_2), \quad Y(0) = 0, \quad (88)$$

$$L = (S_1 Y + S_2)^*. \quad (89)$$

Consider (53). Let $\tilde{T}$ *be as defined in Lemma 4.2. If* $T \geq \tilde{T}$, *there exists* $\tilde{\delta} > 0$ *such that*

$$\int_0^T v^* R v\, dt - \lambda \int_0^T z^* W z\, dt \geq \tilde{\delta} \int_0^T v^* R v\, dt. \quad (90)$$

Proof. From equation (84) of the dual system, we deduce that if $T \geq \tilde{T}$,

$$\int_0^T w^* R w\, dt - \lambda \int_0^T z_1^* W z_1\, dt \geq \delta_1 \int_0^T w^* R w\, dt, \quad (91)$$

for some $\delta_1 > 0$. Since the map $w \to v$ is bounded, there exists $\tilde{\delta} > 0$ such that

$$\int_0^T w^* R w\, dt - \lambda \int_0^T z_1^* W z_1\, dt \geq \tilde{\delta} \int_0^T v^* R v\, dt. \quad (92)$$

From (54), (64), (65), and (92), we get (90). □

The above theorem shows that for the controller defined by (46) and (47), the performance is greater than λ. In fact, from equation (90), we have

$$\inf_{v\neq 0} \frac{\int_0^T v^* R v \, dt}{\int_0^T z^* W z \, dt} > \lambda. \tag{93}$$

5. Summary of Results

The system is given by

$$\dot{x} = A(t)x + B_1(t)u + B_2(t)v, \quad x(0) = 0, \tag{94}$$

$$z = C(t)x + D(t)u + E(t)v, \tag{95}$$

$$y = C_2(t)x + D_2(t)u + E_2(t)v. \tag{96}$$

Let

$$\lambda_{\text{opt}} = \max_u \min_{v\neq 0} \frac{\int_0^T \frac{1}{2} v^* R v \, dt}{\int_0^T \frac{1}{2} z^* W z \, dt}, \tag{97}$$

and let $\lambda < \lambda_{\text{opt}}$. Also, let $W_1, \ldots, W_6$ be defined by

$$z^* W z = x^* W_1 x + 2x^* W_2 u + u^* W_3 u + 2x^* W_4 v + v^* W_5 v + 2u^* W_6 v. \tag{98}$$

Let

$$\tilde{z}_1 = B_2^* \tilde{e} + E_2^* \tilde{u} + E^* \tilde{w}, \tag{99}$$

and let $\tilde{W}_1, \ldots, \tilde{W}_6$ be defined by

$$\tilde{z}_1^* R^{-1} \tilde{z}_1 = \tilde{e}^* \tilde{W}_1 \tilde{e} + 2\tilde{e}^* \tilde{W}_2 \tilde{u} + \tilde{u}^* \tilde{W}_3 \tilde{u} + 2\tilde{e}^* \tilde{W}_4 \tilde{w} + \tilde{w}^* \tilde{W}_5 \tilde{w} + 2\tilde{u}^* \tilde{W}_6 \tilde{w}. \tag{100}$$

We make the following assumptions.

(a) Let the final time T satisfy the assumption in Lemma 4.2.

(b) Assume that for each $t \in [0,T]$, $R - \lambda W_5$ is positive definite and $W_3 + \lambda W_6(R - \lambda W_5)^{-1}W_6^*$ is invertible.

(c) Also assume that for each $t \in [0,T]$, $W^{-1} - \lambda W_5$ is positive definite and $\tilde{W}_3 + \lambda \tilde{W}_6(W^{-1} - \lambda \tilde{W}_5)^{-1}\tilde{W}_6^*$ is invertible.

The relevant controller equations are

$$\Omega = (R - \lambda W_5)^{-1}, \tag{101}$$

$$\Lambda = (W_3 + \lambda W_6 \Omega W_6^*)^{-1}, \tag{102}$$

$$U_1 = \Lambda(B_1^* + \lambda W_6 \Omega B_2^*), \tag{103}$$

$$U_2 = -\Lambda(W_2^* + \lambda W_6 \Omega W_4^*), \tag{104}$$

$$V_1 = \lambda\Omega(-B_2^* + W_6^* U_1), \tag{105}$$

$$V_2 = \lambda\Omega(W_4^* + W_6^* U_2), \tag{106}$$

$$\dot{P} + P(A + B_1U_2 + B_2V_2) + (A^* - W_2U_1 - W_4V_1)P$$
$$+P(B_1U_1 + B_2V_1)P - (W_1 + W_2U_2 + W_4V_2) = 0, \quad P(T) = 0. \tag{107}$$

Note that the above equation is symmetric.

The relevant observer equations are

$$\Phi = (W^{-1} - \lambda\tilde{W}_5)^{-1}, \tag{108}$$

$$\Gamma = (\tilde{W}_3 + \lambda\tilde{W}_6\Phi\tilde{W}_6^*)^{-1}, \tag{109}$$

$$\tilde{A} = A + B_2V_1P + B_2V_2, \tag{110}$$

$$\tilde{B} = -(U_1P + U_2), \tag{111}$$

$$\tilde{C} = C_2 + E_2V_1P + E_2V_2, \tag{112}$$

$$S_1 = \Gamma(\tilde{C} + \lambda\tilde{W}_6\Phi D\tilde{B}), \tag{113}$$

$$S_2 = -\Gamma(\tilde{W}_2^* + \lambda\tilde{W}_6\Phi\tilde{W}_4^*), \tag{114}$$

$$T_1 = \lambda\Phi(-D\tilde{B} + \tilde{W}_6^* S_1), \tag{115}$$

$$T_2 = \lambda\Phi(\tilde{W}_4^* + \tilde{W}_6^* S_2), \tag{116}$$

$$\dot{Y} = (\tilde{A} - \tilde{W}_2 S_1 - \tilde{W}_4 T_1)Y + Y(\tilde{A}^* + \tilde{C}^* S_2 + \tilde{B}^* D^* T_2)$$
$$+Y(\tilde{C}^* S_1 + \tilde{B}^* D^* T_1)Y - (\tilde{W}_1 + \tilde{W}_2 S_2 + \tilde{W}_4 T_2), \quad Y(0) = 0. \quad (117)$$

Note that the above equation is symmetric.

A suboptimal controller is given by

$$\dot{q} = Aq + B_1(U_1 Pq + U_2 q) + B_2(V_1 Pq + V_2 q) + L(\tilde{C}q + D_2 u - y), \quad (118)$$

$$L = (S_1 Y + S_2)^*, \quad (119)$$

$$u = (U_1 P + U_2)q. \quad (120)$$

6. Conclusions

An expression for a suboptimal finite horizon H_∞ controller is derived in a generalized case. The general case has been treated directly without the utilization of transformations. The ouput feedback controller is synthesized via a controller component and an observer component. A summary of all the design equations is given, and these equations are easy to program on a digital computer.

References

[1] Doyle, J. C., Glover, K., Khargonekar, P. P., and Francis, B. A., "State-space solutions to standard H_2 and H_∞ control problems," *IEEE Transactions on Automatic Control*, **34**, 1989, pp. 831–847.

[2] Khargonekar, P. P., "State-space H_∞ Control Theory," in *Mathematical System Theory: The Influence of R. E. Kalman*, edited by A. C. Antoulas, Springer-Verlag, Berlin, 1991.

[3] Khargonekar, P. P., Nagpal, K. M., and Poolla, K. R., "H_∞ control with transients," *SIAM Journal on Control and Optimization*, **29**, 1991, pp. 1373–1393.

[4] RAVI, R., NAGPAL, K. M., AND KHARGONEKAR, P. P., "H^∞ control of linear time-varying systems: A state space approach," *SIAM Journal on Control and Optimization*, **29**, 1991, pp. 1394–1413.

[5] TADMOR, G., "Worst-case design in the time domain: the maximum principle and the standard H_∞ problem," *Mathematics of Control, Signals, and Systems*, **3**, 1990, pp. 301–324.

[6] SUBRAHMANYAM, M. B., *Optimal Control with a Worst-Case Performance Criterion and Applications*, Lecture Notes in Control and Information Sciences, No. 145, Springer-Verlag, Berlin, 1990.

[7] SUBRAHMANYAM, M. B., "Optimal disturbance rejection and performance robustness in linear systems," *Journal of Mathematical Analysis and Applications*, **164**, 1992, pp. 130–150.

[8] SUBRAHMANYAM, M. B., "Synthesis of finite-interval H_∞ controllers by state-space methods," *AIAA Journal of Guidance, Control, and Dynamics*, **13**, 1990, pp. 624–629.

[9] SUBRAHMANYAM, M. B., "Worst-case performance measures for linear control problems," *Proc. IEEE Conference on Decision and Control*, Honolulu, U.S.A., 1990, pp. 2439–2443.

[10] SUBRAHMANYAM, M. B. AND STEINBERG, M., "Model reduction with a finite-interval H_∞ criterion," *Proc. AIAA Guidance, Navigation and Control Conference*, Portland, U.S.A., 1990, pp. 1419–1427.

[11] SAFONOV, M. G., LIMEBEER, D. J. N., AND CHIANG, R. Y., "Simplifying the H_∞ theory via loop-shifting, matrix-pencil and descriptor concepts," *International Journal of Control*, **50**, 1989, pp. 2467–2488.

[12] SUBRAHMANYAM, M. B., "Worst-case optimal control over a finite horizon," *Journal of Mathematical Analysis and Applications*, **171**, 1992, pp. 448–460.

[13] PONTRYAGIN, L. S., BOLTYANSKI, V. G., GAMKRELIDZE, R. V., AND MISCHENKO, E. F., *The Mathematical Theory of Optimal Processes*, Interscience Publishers, New York, 1962.

Chapter 4

Finite Horizon H_∞ with Parameter Variations

Abstract

In this chapter we consider the finite horizon H_∞ performance robustness problem with parameter variations. Assuming the adequacy of linear expressions for performance variation, an iterative procedure is given to synthesize a suboptimal H_∞ controller, which yields the required performance even under parameter variations. As a by-product, an expression for the variation of performance due to parameter variations is given for this specific controller by making use of variational theory. An example which illustrates the methodology is worked out under parameter uncertainties.

1. Introduction

The suboptimal H_∞ problem can been solved via the solution of two algebraic Riccati equations [1]. The finite horizon H_∞ control problem has been treated in [2–6]. In the finite horizon case, expressions for a suboptimal controller have also been derived in a very general time-varying setting [7], without resorting to various transformations, such as the ones given in [8]. In the time-invariant case, the solutions of the dynamic Riccati equations eventually converge to constant matrices. Also, efficient algorithms for the computation of the infimal H_∞ norm have been given in the finite horizon case [9–11]. Our computational experience indicates that for time-invariant systems, the infimal H_∞ norm can be very nearly approached using full state feedback, whereas with output feedback the suboptimal value that gives a viable controller is generally much higher than the infimal value [12].

It is difficult to design a suboptimal controller taking into account parameter variations in the system matrices. It is doubtful whether a noniterative solution exists to the performance robustness problem. The reason for this is the fact that when one designs a suboptimal H_∞ controller at the nominal values of the parameters, the performance of the controller under a variation in the system matrices is *a priori* unknown. In fact, in our treatment here, knowledge of the controller is essential for the computation of the variation of performance under variations in the system matrices. It is tacitly assumed that the controller and observer matrices designed at a nominal point are fixed even under parameter uncertainties, since we have no way to measure the variations in the system matrices in general.

To solve the performance robustness problem, we need to have an idea of the variations in performance. To this end, we derive an expression for the variation in performance for a specific controller in terms of variations in the system matrices. We assume that the accuracy of this linear expression is sufficient for the specified range of parameter variations. If the performance degradation for the range of parameter variations is unsatisfactory, we need to redesign the suboptimal controller.

The chapter is organized as follows. In Section 2, we formulate the finite horizon H_∞ performance robustness problem. Section 3 presents a summary of the equations involved in the design of a suboptimal H_∞ controller. Section 4 presents a procedure for the computation of actual performance of the suboptimal controller. In Section 5, a formula for the degradation in performance due to parameter variations is derived by making use of variational theory. Assuming the adequacy of the linear expression for the degradation in performance derived in Section 5, Section 6 presents an iterative procedure for the design of a suboptimal controller that has adequate performance robustness. An example of the robust control system design that illustrates the usefulness of the theory is given in Section 7.

2. Problem Formulation

Let the n-dimensional linear time-varying system be given by

$$\dot{x} = A(t)x + B_1(t)u + B_2(t)v, \quad x(t_0) = 0, \tag{1}$$

$$z = C(t)x + D(t)u + E(t)v, \tag{2}$$

$$y = C_2(t)x + D_2(t)u + E_2(t)v. \tag{3}$$

Without loss of generality, let $t_0 = 0$. Also let

$$\lambda_{\text{opt}} = \max_{u} \min_{v \neq 0} \frac{\int_0^T \frac{1}{2} v^* R v \, dt}{\int_0^T \frac{1}{2} z^* W z \, dt}, \tag{4}$$

where R and W are assumed to be positive definite and the superscript $*$ denotes a matrix or vector transpose. Computational techniques for the evaluation of λ_{opt} are given in Chapter 2, which is based on [11].

The suboptimal H_∞ problem can be stated as follows. Given $\lambda < \lambda_{\text{opt}}$, find full state and output feedback controllers, if these exist, for which

$$\min_{v \neq 0} \frac{\int_0^T \frac{1}{2} v^* R v \, dt}{\int_0^T \frac{1}{2} z^* W z \, dt} > \lambda. \tag{5}$$

The above problem has been solved in Chapter 3, which is based on [7], and a summary of the design equations will be given in Section 3.

The suboptimal performance robustness problem can be stated as follows. Given $\bar{\lambda} < \lambda_{\text{opt}}$, find state and output feedback controllers, if these exist, for which

$$\min_{v \neq 0} \frac{\int_0^T \frac{1}{2} v^* R v \, dt}{\int_0^T \frac{1}{2} z^* W z \, dt} > \bar{\lambda},$$

even under specified variations in the system matrices in (1)–(3). In this chapter we give an approximate solution to the performance robustness problem by making use of linearized expressions for the variation of performance.

3. Feedback Solutions

In this section, we give equations for a suboptimal H_∞ design. For the details of derivation, see Chapter 3, which is based on [7].

The system is given by

$$\dot{x} = A(t)x + B_1(t)u + B_2(t)v, \quad x(0) = 0, \tag{6}$$

$$z = C(t)x + D(t)u + E(t)v, \tag{7}$$

$$y = C_2(t)x + D_2(t)u + E_2(t)v. \tag{8}$$

Let

$$\lambda_{\text{opt}} = \max_u \min_{v \neq 0} \frac{\int_0^T \frac{1}{2} v^* R v \, dt}{\int_0^T \frac{1}{2} z^* W z \, dt}, \tag{9}$$

and let $\lambda < \lambda_{\text{opt}}$. Also, let $W_1, \ldots, W_6$ be defined by

$$z^* W z = x^* W_1 x + 2x^* W_2 u + u^* W_3 u + 2x^* W_4 v + v^* W_5 v + 2u^* W_6 v. \tag{10}$$

Let

$$\tilde{z}_1 = B_2^* \tilde{e} + E_2^* \tilde{u} + E^* \tilde{w}, \tag{11}$$

and let $\tilde{W}_1, \ldots, \tilde{W}_6$ be defined by

$$\tilde{z}_1^* R^{-1} \tilde{z}_1 = \tilde{e}^* \tilde{W}_1 \tilde{e} + 2\tilde{e}^* \tilde{W}_2 \tilde{u} + \tilde{u}^* \tilde{W}_3 \tilde{u} + 2\tilde{e}^* \tilde{W}_4 \tilde{w} + \tilde{w}^* \tilde{W}_5 \tilde{w} + 2\tilde{u}^* \tilde{W}_6 \tilde{w}. \tag{12}$$

We make the following assumptions.

(a) Let the final time T satisfy the assumption in Lemma 4.2 of Chapter 3.

(b) Assume that for each $t \in [0, T]$, $R - \lambda W_5$ is positive definite and $W_3 + \lambda W_6 (R - \lambda W_5)^{-1} W_6^*$ is invertible.

(c) Also assume that for each $t \in [0, T]$, $W^{-1} - \lambda \tilde{W}_5$ is positive definite and $\tilde{W}_3 + \lambda \tilde{W}_6 (W^{-1} - \lambda \tilde{W}_5)^{-1} \tilde{W}_6^*$ is invertible.

The relevant controller equations are

$$\Omega = (R - \lambda W_5)^{-1}, \tag{13}$$

$$\Lambda = (W_3 + \lambda W_6 \Omega W_6^*)^{-1}, \tag{14}$$

$$U_1 = \Lambda(B_1^* + \lambda W_6 \Omega B_2^*), \tag{15}$$

$$U_2 = -\Lambda(W_2^* + \lambda W_6 \Omega W_4^*), \tag{16}$$

$$V_1 = \lambda\Omega(-B_2^* + W_6^* U_1), \tag{17}$$

$$V_2 = \lambda\Omega(W_4^* + W_6^* U_2), \tag{18}$$

$$\dot{P} + P(A + B_1U_2 + B_2V_2) + (A^* - W_2U_1 - W_4V_1)P$$
$$+P(B_1U_1 + B_2V_1)P - (W_1 + W_2U_2 + W_4V_2) = 0, \quad P(T) = 0. \tag{19}$$

Note that the above equation is symmetric.

The relevant observer equations are

$$\Phi = (W^{-1} - \lambda\tilde{W}_5)^{-1}, \tag{20}$$

$$\Gamma = (\tilde{W}_3 + \lambda\tilde{W}_6\Phi\tilde{W}_6^*)^{-1}, \tag{21}$$

$$\tilde{A} = A + B_2V_1P + B_2V_2, \tag{22}$$

$$\tilde{B} = -(U_1P + U_2), \tag{23}$$

$$\tilde{C} = C_2 + E_2V_1P + E_2V_2, \tag{24}$$

$$S_1 = \Gamma(\tilde{C} + \lambda\tilde{W}_6\Phi D\tilde{B}), \tag{25}$$

$$S_2 = -\Gamma(\tilde{W}_2^* + \lambda\tilde{W}_6\Phi\tilde{W}_4^*), \tag{26}$$

$$T_1 = \lambda\Phi(-D\tilde{B} + \tilde{W}_6^* S_1), \tag{27}$$

$$T_2 = \lambda\Phi(\tilde{W}_4^* + \tilde{W}_6^* S_2), \tag{28}$$

$$\dot{Y} = (\tilde{A} - \tilde{W}_2S_1 - \tilde{W}_4T_1)Y + Y(\tilde{A}^* + \tilde{C}^*S_2 + \tilde{B}^*D^*T_2)$$
$$+Y(\tilde{C}^*S_1 + \tilde{B}^*D^*T_1)Y - (\tilde{W}_1 + \tilde{W}_2S_2 + \tilde{W}_4T_2), \quad Y(0) = 0. \tag{29}$$

Note that the above equation is symmetric.

A suboptimal controller is given by

$$\dot{q} = Aq + B_1(U_1Pq + U_2q) + B_2(V_1Pq + V_2q) + L(\tilde{C}q + D_2u - y), \tag{30}$$

$$L = (S_1Y + S_2)^*, \tag{31}$$

$$u = (U_1P + U_2)q. \tag{32}$$

In the case of time-invariant systems, the solutions of the Riccati equations (19) and (29) eventually converge to constant matrices. Note that the full state feedback controller is simply given by $u = (U_1P + U_2)x$.

4. Computation of Performance

We consider the output feedback case. The full state feedback case is covered by the ensuing analysis as well. The closed loop system is given by

$$\begin{pmatrix} \dot{x} \\ \dot{q} \end{pmatrix} = \begin{pmatrix} A & B_1(U_1P + U_2) \\ -LC_2 & \tilde{U} - LD_2(U_1P + U_2) \end{pmatrix} \begin{pmatrix} x \\ q \end{pmatrix} + \begin{pmatrix} B_2 \\ -LE_2 \end{pmatrix} v, \tag{33}$$

where

$$\tilde{U} = A + B_1(U_1P + U_2) + B_2(V_1P + V_2) + L(\tilde{C} + D_2(U_1P + U_2)). \tag{34}$$

Note that all the matrices can be time-varying. We now specialize to the case where

$$x(0) = q(0) = 0. \tag{35}$$

Let

$$z = Cx + D(U_1P + U_2)q + Ev. \tag{36}$$

For the controller

$$u = (U_1P + U_2)q, \tag{37}$$

the performance is given by

$$\min_{v \neq 0} \frac{\int_0^T \frac{1}{2} v^* R v \, dt}{\int_0^T \frac{1}{2} z^* W z \, dt}. \tag{38}$$

The above value is strictly greater than λ by our design procedure. For time-invariant systems, it is convenient to consider constant solutions of the two dynamic Riccati equations and take the matrices $U_1 P + U_2$ and $S_1 Y + S_2$ to be fixed. In some cases, it is possible that the performance may be below λ because of the disregard for the time dependence. Nevertheless, in such cases the performance can be artificially improved by designing the controller at a higher value of λ than the one specified.

We now describe a procedure to compute the value of (38). Let (33) be rewritten as

$$\dot{x}_s = A_s x_s + B_s v, \quad x_s(0) = 0, \tag{39}$$

where v needs to be chosen to minimize

$$\frac{\int_0^T \frac{1}{2} v^* R v \, dt}{\int_0^T \{\frac{1}{2} x_s^* Q_1 x_s + x_s^* Q_2 v + \frac{1}{2} v^* Q_3 v\} \, dt}. \tag{40}$$

Note that the denominator of (38) can put in the form of the denominator of (40) by virtue of (36).

This problem has been solved previously [13,14], and in addition, conditions guaranteeing the existence of a minimizing v are reported in these references. For the sake of completeness, we present here the details concerning the characterization of a minimizing v and the computation of the performance corresponding to the minimizing v.

In the following theorem we give conditions that are satisfied by an optimal v which minimizes (40) subject to (39).

THEOREM 4.1. *Consider the system given by (39) and (40). Let*

$$\hat{\lambda} = \min_{v \neq 0} \frac{\int_0^T \frac{1}{2} v^* R v \, dt}{\int_0^T \{\frac{1}{2} x_s^* Q_1 x_s + x_s^* Q_2 v + \frac{1}{2} v^* Q_3 v\} \, dt}. \tag{41}$$

Assume that $(R(t) - \hat{\lambda} Q_3(t))^{-1}$ *exists for all* $t \in [0, T]$. *If* (x_s, v) *is optimal, then there exists a nonzero* $\rho(t)$ *such that*

$$\frac{d\rho}{dt} = -A_s^* \rho - \hat{\lambda} Q_1 x_s - \hat{\lambda} Q_2 v, \quad \rho(T) = 0, \tag{42}$$

and

$$v(t) = (R - \hat{\lambda} Q_3)^{-1} \{\hat{\lambda} Q_2^* x_s + B_s^* \rho\}. \tag{43}$$

Proof. If v minimizes (40), then it also minimizes

$$\tilde{J}(v) \stackrel{\text{def}}{=} \int_0^T \frac{1}{2} v^* R v \, dt - \hat{\lambda} \int_0^T \{\frac{1}{2} x_s^* Q_1 x_s + x_s^* Q_2 v + \frac{1}{2} v^* Q_3 v\} \, dt. \tag{44}$$

The theorem now follows from the maximum principle [15]. □

Let

$$\tilde{A} = A_s + \hat{\lambda} B_s (R - \hat{\lambda} Q_3)^{-1} Q_2^*, \tag{45}$$

$$\tilde{B} = B_s (R - \hat{\lambda} Q_3)^{-1} B_s^*, \tag{46}$$

and

$$\tilde{C} = -\hat{\lambda} Q_1 - \hat{\lambda}^2 Q_2 (R - \hat{\lambda} Q_3)^{-1} Q_2^*. \tag{47}$$

The variables satisfy a two point boundary value problem given by

$$\begin{pmatrix} \dot{x}_s \\ \dot{\rho} \end{pmatrix} = \begin{pmatrix} \tilde{A} & \tilde{B} \\ \tilde{C} & -\tilde{A}^* \end{pmatrix} \begin{pmatrix} x_s \\ \rho \end{pmatrix}, \tag{48}$$

with

$$x_s(0) = 0, \quad \rho(T) = 0. \tag{49}$$

We now give a criterion for the estimation of $\hat{\lambda}$. Notice that $\hat{\lambda}$ gives a measure of performance of the H_∞ suboptimal controller under worst-case conditions.

THEOREM 4.2. *Let $\hat{\lambda}$ be the least positive value for which the boundary value problem given by (48) and (49) has a solution (x_s, ρ) with $\int_0^T \{\frac{1}{2}x_s^* Q_1 x_s + x_s^* Q_2 v + \frac{1}{2}v^* Q_3 v\}\,dt > 0$, where $v \stackrel{\text{def}}{=} (R - \hat{\lambda}Q_3)^{-1}\{\hat{\lambda}Q_2^* x_s + B_s^*\rho\}$. Then $\hat{\lambda}$ is the minimum value of (40), x_s is an optimal trajectory, and $v = (R - \hat{\lambda}Q_3)^{-1}\{\hat{\lambda}Q_2^* x_s + B_s^*\rho\}$ is an open loop optimal exogenous input.*

Proof. It is clear from Theorem 4.1 that if $v(t)$ minimizes (40), then it satisfies (48) and (49), with $\hat{\lambda}$ being the minimum value of (40). Now suppose $(x_s, \rho) \neq 0$ satisfies (48) and (49) for some $\hat{\lambda}$. Let $v = (R - \hat{\lambda}Q_3)^{-1}\{\hat{\lambda}Q_2^* x_s + B_s^*\rho\}$. In the following equations $(\cdot\,,\cdot)$ denotes an inner product.

We have

$$\int_0^T \big((R - \hat{\lambda}Q_3)v, v\big)\,dt = \int_0^T (\hat{\lambda}Q_2^* x_s, v)\,dt + \int_0^T (B_s^*\rho, v)\,dt. \tag{50}$$

By equation (39), the second integral of (50) can be written as

$$\int_0^T (B_s^*\rho, v)\,dt = \int_0^T (\rho, B_s v)\,dt = \int_0^T (\rho, \dot{x}_s - A_s x_s)\,dt. \tag{51}$$

An integration by parts and equations (42) and (49) yield

$$\int_0^T (\rho, \dot{x}_s - A_s x_s)\,dt = \hat{\lambda}\int_0^T (Q_1 x_s, x_s)\,dt + \hat{\lambda}\int_0^T (x_s, Q_2 v)\,dt. \tag{52}$$

Substituting (52) in (50), we get

$$\int_0^T v^* R v\,dt = \hat{\lambda}\int_0^T \{x_s^* Q_1 x_s + 2x_s^* Q_2 v + v^* Q_3 v\}\,dt. \tag{53}$$

Thus, the cost associated with v is $\hat{\lambda}$ provided that the right side of (53) is nonzero. Hence, if (x_s, ρ) is a nontrivial solution of the boundary value problem given by (48) and (49) for the smallest parameter $\hat{\lambda} > 0$ with $\int_0^T \{x_s^* Q_1 x_s + 2x_s^* Q_2 v + v^* Q_3 v\}\,dt > 0$, then $\hat{\lambda}$ is the optimal value, and (x_s, ρ) is an optimal pair. □

Note that the boundary value problem (48)–(49) has a solution with a nonvanishing denominator for (40) for at most a countably infinite values of $\hat{\lambda}$. Theorem 4.2 gives a sufficient condition for an open loop exogenous input to be optimal. Theorems 4.1 and 4.2 completely characterize the open loop worst-case exogenous input.

Making use of the transition matrix, the solution of (48) may be expressed as

$$\begin{pmatrix} x_s(t) \\ \rho(t) \end{pmatrix} = \begin{pmatrix} \Phi_{11}(t,0) & \Phi_{12}(t,0) \\ \Phi_{21}(t,0) & \Phi_{22}(t,0) \end{pmatrix} \begin{pmatrix} x_s(0) \\ \rho(0) \end{pmatrix}. \tag{54}$$

Equation (49) yields

$$\Phi_{12}(T,0)\rho(0) = x_s(T), \tag{55}$$

$$\Phi_{22}(T,0)\rho(0) = 0. \tag{56}$$

In view of (56) and (48)–(49), we have $\det(\Phi_{22}(T,0)) = 0$ if and only if the solution (x_s, ρ) of (48)–(49) is not identically zero. Thus, we need the least positive $\hat{\lambda}$ which makes both $\det(\Phi_{22}(T,0)) = 0$ and the denominator of (40) positive. This can be very easily obtained by doing a search with $\hat{\lambda}$ over an interval on which there is a change in the sign of the determinant.

We found the following algorithm to be numerically more stable, since numbers of lesser magnitude are involved in the computation of the transition matrices in (57). We have

$$\begin{pmatrix} x_s(T) \\ \rho(T) \end{pmatrix} = \Phi(T,T/2)\Phi(T/2,0) \begin{pmatrix} x_s(0) \\ \rho(0) \end{pmatrix}. \tag{57}$$

Let

$$\Phi^{-1}(T,T/2) = \begin{pmatrix} \xi_{11} & \xi_{12} \\ \xi_{21} & \xi_{22} \end{pmatrix},$$

and

$$\Phi(T/2,0) = \begin{pmatrix} \nu_{11} & \nu_{12} \\ \nu_{21} & \nu_{22} \end{pmatrix}.$$

Making use of $x_s(0) = \rho(T) = 0$, we have

$$\nu_{12}\rho(0) = \xi_{11}x_s(T), \tag{58}$$

$$\nu_{22}\rho(0) = \xi_{21}x_s(T). \tag{59}$$

Thus

$$\det\begin{pmatrix} \nu_{12} & \xi_{11} \\ \nu_{22} & \xi_{21} \end{pmatrix} = 0. \tag{60}$$

Thus we need the least positive $\hat{\lambda}$ which makes the above determinant zero.

5. Performance Variation

In this section we develop a formula for the variation of $\hat{\lambda}$ when there are parameter variations in the system matrices. The system is given by

$$\dot{x}_s = A_s x_s + B_s v, \quad x_s(0) = 0, \tag{61}$$

with v chosen to minimize

$$\frac{\int_0^T \frac{1}{2}v^* R v \, dt}{\int_0^T \{\frac{1}{2}x_s^* Q_1 x_s + x_s^* Q_2 v + \frac{1}{2}v^* Q_3 v\}}. \tag{62}$$

From the theory of Section 4, we have a boundary value problem given by

$$\dot{x}_s = \tilde{A}x_s + \tilde{B}\rho, \tag{63}$$

$$\dot{\rho} = \tilde{C}x_s - \tilde{A}^*\rho, \tag{64}$$

with

$$x_s(0) = 0, \quad \rho(T) = 0. \tag{65}$$

Note that $\hat{\lambda}$ is the minimal value of (62). Because of variations in the matrices of the original system (1)–(3), there will be corresponding variations in the matrices $\tilde{A}(t), \tilde{B}(t)$, and $\tilde{C}(t)$. Let the elemental dependent variations in $\tilde{A}, \tilde{B}$, and $\tilde{C}$ be denoted by $\delta\tilde{A}, \delta\tilde{B}$, and $\delta\tilde{C}$, respectively. For simplicity of notation, we will derive the variation in performance in terms of $\delta\tilde{A}, \delta\tilde{B}$, and $\delta\tilde{C}$. We denote by $\delta\hat{\lambda}$ the variation in $\hat{\lambda}$ due to the parameter variations.

Let

$$\Upsilon = (R - \hat{\lambda}Q_3)^{-1}. \tag{66}$$

We have

$$\delta\tilde{A} = I_1 + \delta\hat{\lambda}J_1, \tag{67}$$

$$\delta\tilde{B} = I_2 + \delta\hat{\lambda}J_2, \tag{68}$$

$$\delta\tilde{C} = I_3 + \delta\hat{\lambda}J_3, \tag{69}$$

where

$$I_1 = \delta A_s + \hat{\lambda}\delta B_s\Upsilon Q_2^* + \hat{\lambda}B_s\Upsilon\big(\delta R - \hat{\lambda}\delta Q_3\big)\Upsilon Q_2^* + \hat{\lambda}B_s\Upsilon\delta Q_2^*, \tag{70}$$

$$J_1 = -\hat{\lambda}B_s\Upsilon Q_3\Upsilon Q_2^* + B_s\Upsilon Q_2^*, \tag{71}$$

$$I_2 = \delta B_s\Upsilon B_s^* + B_s\Upsilon\big(\delta R - \hat{\lambda}\delta Q_3\big)\Upsilon B_s^* + B_s\Upsilon\delta B_s^*, \tag{72}$$

$$J_2 = -B_s\Upsilon Q_3\Upsilon B_s^*, \tag{73}$$

$$I_3 = -\hat{\lambda}\delta Q_1 - \hat{\lambda}^2\delta Q_2\Upsilon Q_2^* - \hat{\lambda}^2 Q_2\Upsilon\big(\delta R - \hat{\lambda}\delta Q_3\big)\Upsilon Q_2^* - \hat{\lambda}^2 Q_2\Upsilon\delta Q_2^*, \tag{74}$$

$$J_3 = -Q_1 - 2\hat{\lambda}Q_2\Upsilon Q_2^* + \hat{\lambda}^2 Q_2\Upsilon Q_3\Upsilon Q_2^*. \tag{75}$$

Let x_1 and ρ_1 represent the variations in x_s and ρ due to $\delta\tilde{A}, \delta\tilde{B}$, and $\delta\tilde{C}$. From (63)–(65), we have the following equations that are satisfied by x_1 and ρ_1.

$$\dot{x}_1 = \tilde{A}x_1 + \tilde{B}\rho_1 + (I_1 + \delta\hat{\lambda}J_1)x_s + (I_2 + \delta\hat{\lambda}J_2)\rho, \tag{76}$$

$$\dot{\rho}_1 = \tilde{C}x_1 - \tilde{A}^*\rho_1 + (I_3 + \delta\hat{\lambda}J_3)x_s - (I_1 + \delta\hat{\lambda}J_1)^*\rho, \tag{77}$$

$$x_1(0) = 0, \quad \rho_1(T) = 0. \tag{78}$$

THEOREM 5.1. *The variation* $\delta\hat{\lambda}$ *in performance is given by*

$$\delta\hat{\lambda} = \frac{-\int_0^T \{x_s^* I_1^* \rho + \frac{1}{2}\rho^* I_2 \rho - \frac{1}{2} x_s^* I_3 x_s\}\, dt}{\int_0^T \{\frac{1}{2} x_s^* Q_1 x_s + x_s^* Q_2 v + \frac{1}{2} v^* Q_3 v\}\, dt}, \tag{79}$$

where $v = \Upsilon\{B_s^* \rho + \hat{\lambda} Q_2^* x_s\}$.

Proof. From (77) we obtain

$$\int_0^T x_s^* \dot{\rho}_1 \, dt = \int_0^T x_s^* \tilde{C} x_1 \, dt - \int_0^T x_s^* \tilde{A}^* \rho_1 \, dt + \int_0^T x_s^* (I_3 + \delta\hat{\lambda} J_3) x_s \, dt - \int_0^T x_s^* (I_1 + \delta\hat{\lambda} J_1)^* \rho \, dt. \tag{80}$$

Integrating the left side of (80) by parts and making use of (63) and (65), we obtain

$$-\int_0^T \rho_1^* \hat{B} \rho \, dt = \int_0^T x_s^* \tilde{C} x_1 \, dt + \int_0^T x_s^* (I_3 + \delta\hat{\lambda} J_3) x_s \, dt - \int_0^T x_s^* (I_1 + \delta\hat{\lambda} J_1)^* \rho \, dt. \tag{81}$$

By (64), the first integral on the right side of (81) is written as

$$\int_0^T x_s^* \tilde{C} x_1 \, dt = \int_0^T (\dot{\rho} + \tilde{A}^* \rho)^* x_1 \, dt. \tag{82}$$

An integration by parts and equations (76) and (78) yield

$$\int_0^T x_s^* \tilde{C} x_1 \, dt = -\int_0^T \rho^* \tilde{B} \rho_1 \, dt - \int_0^T \rho^* (I_1 + \delta\hat{\lambda} J_1) x_s \, dt - \int_0^T \rho^* (I_2 + \delta\hat{\lambda} J_2) x_s \, dt. \tag{83}$$

Substituting (83) in (81) and simplifying, we obtain

$$\delta\hat{\lambda} = \frac{-\int_0^T \{2 x_s^* I_1^* \rho + \rho^* I_2 \rho - x_s^* I_3 x_s\}\, dt}{\int_0^T \{2 x_s^* J_1^* \rho + \rho^* J_2 \rho - x_s^* J_3 x_s\}\, dt}. \tag{84}$$

It can be easily shown (see Lemma 1 of [13], for example) that the denominator of (84) is equal to $\int_0^T \{x_s^* Q_1 x_s + 2 x_s^* Q_2 v + v^* Q_3 v\}\, dt$. □

6. Performance Robustness Problem Solution

We now give an iterative procedure to approximately solve the performance robustness problem posed in Section 2. The equations are given by

$$\dot{x} = A(t)x + B_1(t)u + B_2(t)v, \quad x(0) = 0, \tag{85}$$

$$z = C(t)x + D(t)u + E(t)v, \tag{86}$$

$$y = C_2(t)x + D_2(t)u + E_2(t)v, \tag{87}$$

and

$$\lambda_{\text{opt}} = \max_u \min_{v \neq 0} \frac{\int_0^T \frac{1}{2} v^* R v \, dt}{\int_0^T \frac{1}{2} z^* W z \, dt}. \tag{88}$$

The suboptimal performance robustness problem is to find a controller, if one exists, for which

$$\min_{v \neq 0} \frac{\int_0^T \frac{1}{2} v^* R v \, dt}{\int_0^T \frac{1}{2} z^* W z \, dt} > \bar{\lambda}, \quad \bar{\lambda} < \lambda_{\text{opt}}, \tag{89}$$

under variations in the system matrices involved in (85)–(87).

The iterative steps are as follows.

1. Set $\lambda = \lambda_i$, $\lambda_i < \lambda_{\text{opt}}$, and design a suboptimal H_∞ controller based on the design equations of Section 3.
2. For this controller, let

$$\hat{\lambda} = \min_{v \neq 0} \frac{\int_0^T \frac{1}{2} v^* R v \, dt}{\int_0^T \{\frac{1}{2} x_s^* Q_1 x_s + x_s^* Q_2 v + \frac{1}{2} v^* Q_3 v\} \, dt}. \tag{90}$$

This can be computed using the results of Section 4. Note that $\hat{\lambda} > \lambda$ in many problems, even when the time dependence of the matrices $U_1P + U_2$ and $S_1Y + S_2$ is ignored. If $\hat{\lambda} \leq \bar{\lambda}$, increase the value of λ and go through Steps 1 and 2 until $\hat{\lambda} > \bar{\lambda}$.

3. For the range of allowable variations of the matrices involved in (85)–(87), find the worst-case value of $\delta\hat{\lambda}$ using (79). Note that (79) is linear in the parameter variations. Thus, let $\delta = \min \delta\hat{\lambda}$. The controller of Step 1 solves the performance robustness problem in case $\hat{\lambda} + \delta > \bar{\lambda}$. If this is the case, stop the iteration and use the controller of Step 1.
4. If $\hat{\lambda} + \delta \leq \bar{\lambda}$, choose a suitable λ_{i+1}, $\lambda_{i+1} < \lambda_{\text{opt}}$, and set $\lambda = \lambda_{i+1}$. Go back to Step 1. Note that λ_{i+1} may be bigger or smaller than λ_i, depending on the specific problem being solved.

Alternately we can get $\hat{\lambda}$ and $\delta = \min \delta\hat{\lambda}$ for a range of values of λ and then pick a value of λ that solves the performance robustness problem. In case we cannot find a value of λ which satisfies the performance robustness requirement, then the requirement is too stringent and the performance robustness problem does not have a solution. The requirement can be relaxed either by lowering $\bar{\lambda}$ or by suitably shrinking the range of allowable parameter variations.

Since only linear expressions are utilized in the computation for the worst-case variation in performance, it is clear that the above approach yields only an approximate procedure for the solution of the performance robustness problem.

7. An Example

We now illustrate the theory with an example. The system equations are given by (85)–(88) with

$$A = \begin{pmatrix} 1 & 1 \\ 0 & -1 \end{pmatrix}, \quad B_1 = \begin{pmatrix} 1 \\ 0 \end{pmatrix}, \quad B_2 = \begin{pmatrix} 0 & 0 \\ 1 & 0 \end{pmatrix},$$

$$C = \begin{pmatrix} 1 & 0 \\ 0 & 0 \end{pmatrix}, \quad D = \begin{pmatrix} 0 \\ 1 \end{pmatrix}, \quad E = \begin{pmatrix} 0 & 0 \\ 0 & 0 \end{pmatrix},$$

$$C_2 = (1 \quad 0), \quad D_2 = 0, \quad E_2 = (0 \quad 1), \quad R = 100 I_2, \quad W = I_2,$$

where I_2 is the two-dimensional identity matrix. Let

$$a_1 = \delta A(1,1),$$

$$a_2 = \delta B_1(1,1),$$

$$a_3 = \delta B_2(2,1).$$

Assume that $a_i \in [-0.05, 0.05], i = 1,2,3$. Given a $\bar{\lambda}$, the problem is to find a controller for which

$$\min_{v \neq 0} \frac{\int_0^5 v^* R v \, dt}{\int_0^5 z^* W z \, dt} > \bar{\lambda} \tag{91}$$

under the above variations in $A(1,1), B_1(1,1)$, and $B_2(2,1)$.

We can find λ_{opt} given by (4) using the technique given in Chapter 2. For this example, $\lambda_{\text{opt}} = 261.4016$. A suboptimal full state feedback controller can be designed choosing λ arbitrarily close to λ_{opt}. However, to obtain a viable suboptimal output feedback design, λ needs to be chosen much lower than λ_{opt}. Let us form Table 1, which gives the performance $\hat{\lambda}$ and the worst-case variation δ for various values of λ.

TABLE 1

λ	$\hat{\lambda}$	$\delta = \min \delta\hat{\lambda}$	$\hat{\lambda} + \delta$
1	5.4601	−1.8978	3.5623
2	5.8783	−1.9811	3.8972
3	6.3303	−2.0659	4.2644
4	6.8179	−2.1519	4.6660
5	7.3431	−2.2388	5.1043
6	7.9079	−2.3262	5.5817
7	8.5140	−2.4132	6.1008
8	9.1630	−2.4988	6.6642
9	9.8562	−2.5809	7.2753
10	10.5944	−2.6551	7.9393

We now explain the procedure to obtain a particular row of the table. For example, for $\lambda = 3$, we find the output feedback controller using the equations given in Section 3. Note that we ignore the time dependence of the solutions of (19) and (29) and only consider the constant solutions. For $\lambda = 3$,

$$U_1P + U_2 = (\,-2.4249 \quad -1.0062\,),$$

$$V_1P + V_2 = \begin{pmatrix} 0.0302 & 0.0151 \\ 0 & 0 \end{pmatrix},$$

$$L = \begin{pmatrix} -2.7524 \\ -0.3047 \end{pmatrix}.$$

The closed loop system (33) is given by

$$\begin{pmatrix} \dot{x} \\ \dot{q} \end{pmatrix} = \begin{pmatrix} 1 & 1 & -2.4249 & -1.0062 \\ 0 & -1 & 0 & 0 \\ 2.7524 & 0 & -4.1773 & -0.0062 \\ 0.3047 & 0 & -0.2745 & -0.9849 \end{pmatrix} \begin{pmatrix} x \\ q \end{pmatrix} + \begin{pmatrix} 0 & 0 \\ 1 & 0 \\ 0 & 2.7524 \\ 0 & 0.3047 \end{pmatrix} v, \tag{92}$$

with $x(0) = q(0) = 0$. Going through the procedure given in Section 4, the actual performance $\hat{\lambda}$ of the output feedback controller

$$u = (U_1P + U_2)q \tag{93}$$

can be obtained as the first positive value of $\hat{\lambda}$ which satisfies (60). This value is $\hat{\lambda} = 6.3303$. Using (58)–(59) and selecting, for example, the first component of $\rho(0)$ as 1, we can obtain $\rho(0)$. Thus, we can solve the initial value problem (63)–(64) with $x_s(0) = 0$ and the value of $\rho(0)$ as obtained above.

Let $\rho = (\,\rho_u^* \quad \rho_l^*\,)^*$, where ρ_u and ρ_l have the same dimension. For this particular example, (79) becomes

$$\delta\hat{\lambda} = \frac{-\int_0^5 N(t)\,dt}{\int_0^5 x_s^* Q_1 x_s\,dt}, \tag{94}$$

where

$$N = 2\big[(\rho_u, \delta A x) + (\rho_u, \delta B_1 U_1 P q) + 0.01\big(\rho_u, \delta B_2 (B_2^* \rho_u - (LE_2)^* \rho_l)\big)\big]. \tag{95}$$

Note that since the controller is assumed to be fixed, $\tilde{U}$ and L in (33) will be fixed, and thus δA_s and δB_s in (70)–(75) are given by

$$\delta A_s = \begin{pmatrix} \delta A & \delta B_1(U_1 P + U2) \\ 0 & 0 \end{pmatrix}, \quad \delta B_s = \begin{pmatrix} \delta B_2 \\ 0 \end{pmatrix}. \tag{96}$$

Thus, the numerator of (94) is minimized if

$$\delta A = \delta A^{\max} \circ \operatorname{sign}(\rho_u x^*), \tag{97}$$

$$\delta B_1 = \delta B_1^{\max} \circ \operatorname{sign}(\rho_u (U_1 P q)^*), \tag{98}$$

$$\delta B_2 = \delta B_2^{\max} \circ \operatorname{sign}\big(\rho_u (B_2^* \rho_u - (L E_2)^* \rho_l)^*\big). \tag{99}$$

Here in (97)–(99), the equality is element-wise. Thus, for example, in (97),

$$\delta A_{ij} = \delta A_{ij}^{\max} \operatorname{sign}\big((\rho_u x^*)_{ij}\big).$$

Substituting (97)–(99) in (94), we get the worst-case value of $\delta\hat{\lambda}$. From Table 1, $\delta = \min \delta\hat{\lambda} = -2.0659$. Assuming that the accuracy of linear expressions is acceptable, $\hat{\lambda} + \delta$ in Table 1 gives the greatest lower bound of the actual performance of the controller under parameter variations. This value is 4.2644.

Going back to (91), if $\bar{\lambda} = 5$, looking at Table 1, λ can be 5 or higher. If $\bar{\lambda} = 6$, the controller design can be performed with $\lambda = 7$, or with a higher value of λ. If $\bar{\lambda} = 7$, we can select λ at 9 or higher.

8. Conclusions

In this chapter an approach is presented for the solution of the finite horizon H_∞ performance robustness problem under parameter variations. In the case of the proposed suboptimal controller, a linear expression for the degradation of performance is given in terms of variations in the system matrices. An example which illustrates the methodology is also given.

References

[1] Doyle, J. C., Glover, K., Khargonekar, P. P., and Francis, B. A., "State-space solutions to standard H_2 and H_∞ control problems," *IEEE Transactions on Automatic Control*, **34**, 1989, pp. 831–847.

[2] Tadmor, G., "Worst-case design in the time-domain: the maximum principle and the standard H_∞ problem," *Mathematics of Control, Signals, and Systems*, **3**, 1990, pp. 301–324.

[3] Limebeer, D. J. N., Anderson, B. D. O., Khargonekar, P. P., and Green, M., "A game theoretic approach to H_∞ for time-varying systems," *SIAM Journal on Control and Optimization*, **30**, 1992, pp. 262–283.

[4] Stoorvogel, A.A. and Trentelman, H.L., "The finite horizon singular time-varying H_∞ control problem with dynamic measurement feedback," *COSOR memorandum 89-33*, Eindhoven University of Technology, 1989.

[5] Trentelman, H.L. and Stoorvogel, A.A., "Completion of the squares in the finite horizon H_∞ control problem by measurement feedback," in *New Trends in System Theory*, edited by G. Conte, A. M. Perdon, and B. Wyman, Proc. of the Università di Genova-The Ohio State University Joint Conference, Birkhäuser, Boston, 1991.

[6] Stoorvogel, A.A., *The H_∞ Control Problem: A State Space Approach*, Prentice Hall International, Englewood Cliffs, N.J., 1992.

[7] Subrahmanyam, M. B., "General formulae for suboptimal H_∞ control over a finite horizon," *International Journal of Control*, **57**, 1993, pp. 365–375.

[8] SAFONOV, M. G., LIMEBEER, D. J. N., AND CHIANG, R. Y., "Simplifying the H_∞ theory via loop-shifting, matrix-pencil and descriptor concepts," *International Journal of Control*, **50**, 1989, pp. 2467–2488.

[9] SUBRAHMANYAM, M. B., "Computation of infimal H_∞ norm over a finite horizon," *International Journal of Robust and Nonlinear Control*, **2**, 1992, pp. 49–61.

[10] SUBRAHMANYAM, M. B., "Worst-case optimal control over a finite horizon," *Journal of Mathematical Analysis and Applications*, **171**, 1992, pp. 448–460.

[11] SUBRAHMANYAM, M. B., "Synthesis of suboptimal H_∞ controllers over a finite horizon," *International Journal of Control*, **57**, 1993, pp. 351–364.

[12] SUBRAHMANYAM, M. B., "H_∞ design of the F/A-18A automatic carrier landing system," *Tech. Rep. NAWCADWAR-92043-60*, Naval Air Warfare Center, Aircraft Division, Warminster, PA 18974-0591, 1992.

[13] SUBRAHMANYAM, M. B., "Optimal disturbance rejection and performance robustness in linear systems," *Journal of Mathematical Analysis and Applications*, **164**, 1992, pp. 130–150.

[14] SUBRAHMANYAM, M. B., *Optimal Control with a Worst-Case Performance Criterion and Applications*, Lecture Notes in Control and Information Sciences, No. 145, Springer-Verlag, Berlin, 1990.

[15] PONTRYAGIN, L. S., BOLTYANSKI, V. G., GAMKRELIDZE, R. V., AND MISCHENKO, E. F., *The Mathematical Theory of Optimal Processes*, Interscience Publishers, New York, 1962.

Chapter 5

A General Minimization Problem with Application to Performance Robustness in Finite Horizon H_∞

Abstract

In this chapter we treat a general minimization problem on a finite horizon in which the cost functional is a quotient of two definite integrals. An existence theorem is given for the minimization of the cost functional, and necessary conditions that need to be satisfied by the minimizer are stated. Also, a condition is given for the evaluation of the minimum value of the cost functional. The results are shown to have application to the finite horizon H_∞ performance robustness problem. An expression for the variation of the performance in terms of variations in the system matrices is developed.

1. Introduction

In this chapter we consider a general minimization problem for linear systems that involves the minimization of a quotient of definite integrals. Minimization problems involving such functionals have been treated previously [1–3] in the context of integral inequalities. In the present case the solution of the minimization problem is useful for the computation of the finite horizon H_∞ norm of the closed loop system corresponding to a given controller. Also, an expression for the variation of performance due to variations in the system matrices can be developed for this controller. In particular, if we choose a suboptimal H_∞ controller, these expressions are useful in guaranteeing that the performance does not fall below a certain level, even when there are variations in the plant matrices. Of course, this assumes that we are satisfied with the accuracy of linear expressions in computing performance variations.

The present chapter is organized as follows. Section 2 formulates a general minimization problem which will be shown later to have application to the H_∞ performance robustness problem. Existence results for a minimizing input are also given in Section 2. In Section 3, necessary and sufficient conditions for the minimization of the functional involved are given. These conditions will be useful in computing the H_∞ norm of the closed loop system for a given control. Section 4 gives an expression for the variation of the functional in terms of parameter variations. For a given controller, this linear expression can be used to compute the degradation in performance when parameters are perturbed from the nominal values. The results of Sections 2–4 are applied to the finite horizon H_∞ performance robustness problem in Section 5.

2. Existence of a Minimizer

Consider the system given by

$$\dot{\zeta} = M(t)\zeta + N(t)v, \qquad \zeta(0) = 0. \tag{1}$$

The functional to be minimized is

$$F(\zeta, v) = \frac{\int_0^T \phi_1(v,t)\,dt}{\int_0^T \phi_2(\zeta, v, t)\,dt}, \tag{2}$$

where v is a measurable input. We make the following assumptions.

(i) $M(t)$ and $N(t)$ are continuous on $[0,T]$.

(ii) For $i = 1, 2$, ϕ_i is continuous in its arguments.

(iii) ϕ_1 is convex in v for each t. Also, let $\phi_2(\zeta, v, t) = \phi_2^1(\zeta, t) + \phi_2^2(v, t)$.

(iv) Admissible v are measurable functions such that $|\int_0^T \phi_1(v,t)\,dt| < \infty$.

(v) For almost all $t \in [0,T]$, $\phi_1(v,t) \geq a|v|^p$, $a > 0$, $p > 1$, and $\phi_2(\zeta, v, t) > 0$ unless $(\zeta, v) = 0$.

(vi) For each $K < \infty$, there is an integrable $g_K(t)$ such that for each $\|v\|_p \leq K$, $|\phi_2^1(\zeta, t)| \leq g_K(t)$ a.e. on $[0, T]$, where ζ is the response of (1).

(vii) Let

$$\hat{\lambda} = \inf_{v \neq 0} \frac{\int_0^T \phi_1(v, t)\, dt}{\int_0^T \phi_2(\zeta, v, t)\, dt}.$$

Assume that $\phi_1(v, t) - \hat{\lambda}\phi_2^2(v, t)$ is convex for each fixed t and further $\phi_1(v, t) - \hat{\lambda}\phi_2^2(v, t) \geq \alpha\phi_2^2(v, t)$ for some $\alpha > 0$.

(viii) There exists $k > 0$ such that for any $c \geq 0$,

$$\phi_1(cv, t) = c^k \phi_1(v, t), \qquad \phi_2(c\zeta, cv, t) = c^k \phi_2(\zeta, v, t).$$

(ix) There is an admissible v such that $0 < \int_0^T \phi_2(\zeta, v, t)\, dt < \infty$.

In this chapter, $L_p^m(0, T)$ denotes the Banach space of m-dimensional measurable functions on $[0, T]$ with the norm $\|v\|_p = \left[\int_0^T |v(t)|^p\, dt\right]^{1/p}$, where $|\cdot|$ denotes the finite-dimensional Euclidean norm. Also, $C(0, T)$ denotes the space of all continuous $\zeta(t)$ such that $\|\zeta\| = \sup_{[0,T]} |\zeta(t)|$.

THEOREM 2.1. *Consider (1) and (2) along with assumptions (i)–(ix). Then there exists an admissible v that minimizes (2).*

Proof. By the assumptions, it is sufficient to exhibit a minimizing v among all v for which $\int_0^T \phi_2(\zeta, v, t)\, dt = 1$. By assumptions (v), (vii), and (viii), we have $\inf_v \int_0^T \phi_1(v, t)\, dt = \hat{\lambda}$ subject to $\int_0^T \phi_2(\zeta, v, t)\, dt = 1$. Let $\{v_n\}$ be such that $\lim_{n \to \infty} \int_0^T \phi_1(v_n, t)\, dt = \hat{\lambda}$ with $\int_0^T \phi_2(\zeta_n, v_n, t)\, dt = 1$ for all n. Furthermore, the sequence $\{v_n\}$ can be selected such that $\left\{\int_0^T \phi_2^1(\zeta_n, t)\, dt\right\}$ and $\left\{\int_0^T \phi_2^2(v_n, t)\, dt\right\}$ are convergent. By assumption (v), $\{v_n\}$ forms a bounded sequence in $L_p^m(0, T)$, and hence a subsequence, still denoted by $\{v_n\}$, converges weakly to some $v_0 \in L_p^m(0, T)$. Let ζ_0 be the response of

(1) to v_0. By assumption (i) and by the weak convergence of $\{v_n\}$, $\{\zeta_n\}$ converges to ζ_0 pointwise. By assumption (vi) and the Lebesgue dominated convergence theorem, $\lim_{n\to\infty}\int_0^T \phi_2^1(\zeta_n,t)\,dt = \int_0^T \phi_2^1(\zeta_0,t)\,dt$. We have by assumption (iii) and the weak convergence of $\{v_n\}$ to v_0 (see [4, p. 209]), $\int_0^T \phi_1(v_0,t)\,dt \le \liminf_{n\to\infty}\int_0^T \phi_1(v_n,t)\,dt$.

We claim that $\int_0^T \phi_2(\zeta_0,v_0,t)\,dt > 0$. If not, by assumption (v), $v_0 = 0$ almost everywhere and $\zeta_0 = 0$ on $[0,T]$. Thus $\lim_{n\to\infty}\int_0^T \phi_2^1(\zeta_n,t)\,dt = 0$. Since $\lim_{n\to\infty}\int_0^T \phi_2(\zeta_n,v_n,t)\,dt = 1$, this implies that $\lim_{n\to\infty}\int_0^T \phi_2^2(v_n,t)\,dt = 1$. By assumption (vii), $\phi_1(v_n,t) - \hat{\lambda}\phi_2^2(v_n,t) \ge \alpha\phi_2^2(v_n,t)$. Integrating both sides from 0 to T and letting $n\to\infty$, we get the contradiction that $\hat{\lambda}-\hat{\lambda} \ge \alpha > 0$. Thus $\int_0^T \phi_2(\zeta_0,v_0,t)\,dt > 0$.

We now show that

$$\frac{\int_0^T \phi_1(v_0,t)\,dt}{\int_0^T \phi_2(\zeta_0,v_0,t)\,dt} = \hat{\lambda}.$$

Since the left side is by the definition of $\hat{\lambda}$ the larger quantity, it is sufficient to show that

$$\frac{\int_0^T \phi_1(v_0,t)\,dt}{\int_0^T \phi_2(\zeta_0,v_0,t)\,dt} \le \hat{\lambda}.$$

Since $\lim_{n\to\infty}\int_0^T \phi_1(v_n,t)\,dt = \hat{\lambda}$, it is enough to show that

$$\lim_{n\to\infty}\int_0^T \phi_1(v_n,t)\,dt - \frac{\int_0^T \phi_1(v_0,t)\,dt}{\int_0^T \phi_2(\zeta_0,v_0,t)\,dt} \ge 0.$$

Since $\lim_{n\to\infty}\int_0^T \phi_2(\zeta_n,v_n,t)\,dt = 1$, the left side of the above inequality is

$$\lim_{n\to\infty}\int_0^T \phi_1(v_n,t)\,dt - \frac{\int_0^T \phi_1(v_0,t)\,dt}{1-\lim_{n\to\infty}\int_0^T \phi_2^2(v_n,t)\,dt + \int_0^T \phi_2^2(v_0,t)\,dt}.$$

The numerator of the above quantity is

$$\hat{\lambda} - \lim_{n\to\infty}\hat{\lambda}\int_0^T \phi_2^2(v_n,t)\,dt + \hat{\lambda}\int_0^T \phi_2^2(v_0,t)\,dt - \int_0^T \phi_1(v_0,t)\,dt.$$

This, in turn, is equal to

$$\lim_{n\to\infty}\int_0^T [\phi_1(v_n,t)-\hat{\lambda}\phi_2^2(v_n,t)]\,dt-\int_0^T [\phi_1(v_0,t)-\hat{\lambda}\phi_2^2(v_0,t)]\,dt. \quad (3)$$

By assumption (vii), $\phi_1(v,t)-\hat{\lambda}\phi_2^2(v,t)$ is convex for each fixed t. Thus

$$\int_0^T [\phi_1(v_0,t)-\hat{\lambda}\phi_2^2(v_0,t)]\,dt \le \liminf_{n\to\infty}\int_0^T [\phi_1(v_n,t)-\hat{\lambda}\phi_2^2(v_n,t)]\,dt.$$

Thus the quantity given by (3) is nonnegative. The proof of the theorem is hence complete. □

3. Characterization of v_0 and $\hat{\lambda}$

In this section we derive a boundary value problem to be satisfied by a minimizing v_0. This boundary value problem will have $\hat{\lambda}$ as a parameter. We will characterize $\hat{\lambda}$ as the least positive value for which the boundary value problem has a nontrivial solution. We consider (1) and (2) under the following assumptions.

(i) $M(t)$ and $N(t)$ are continuous on $[0,T]$.

(ii) Admissible v are measurable functions such that $|\int_0^T \phi_1(v,t)\,dt| < \infty$.

(iii) $\phi_1(v,t)\ge a|v|^p$, $a>0$, $p>1$, and $\phi_2(\zeta,v,t)\ge 0$ for almost all $t\in[0,T]$.

(iv) Let ϕ_1 be continuously differentiable in v and measurable in t, and let ϕ_2 be continuously differentiable in (ζ,v) and measurable in t.

(v) $\nabla_v\phi_1\in L_q^m(0,T)$ for all $v\in L_p^m(0,T)$, $1/p+1/q=1$.

(vi) Let $\nabla_v\phi_1(v,t)$ and $(\nabla_\zeta\phi_2,\nabla_v\phi_2)(\zeta,v,t)$ be bounded for all bounded (ζ,v).

(vii) (x_0,v_0) minimizes (2) subject to (1)

$$\Rightarrow 0<\int_0^T \phi_1(v_0,t)\,dt<\infty,\quad 0<\int_0^T \phi_2(\zeta_0,v_0,t)\,dt<\infty.$$

We now state the necessary conditions satisfied by an optimal v_0. Note that the superscript $*$ in the case of a matrix or a vector denotes its transpose.

THEOREM 3.1. *Consider (1) and (2) along with the above assumptions. Suppose v_0 minimizes (2). Then there exists an adjoint response $\psi(t)$ such that*

$$\frac{d\psi}{dt} = -M^*(t)\psi - \hat{\lambda}\nabla_\zeta \phi_2(\zeta_0, v_0, t), \qquad \psi(T) = 0, \tag{4}$$

where

$$\hat{\lambda} = \frac{\int_0^T \phi_1(v_0,t)\,dt}{\int_0^T \phi_2(\zeta_0, v_0, t)\,dt} \tag{5}$$

and

$$\nabla_v\phi_1(v_0,t) - \hat{\lambda}\nabla_v\phi_2(\zeta_0,v_0,t) - N^*(t)\psi(t) = 0 \text{ a.e. } on\ [0,T]. \tag{6}$$

Proof. Let $E = C(0,T) \times L_p^m(0,T)$. We have an equality constraint in E given by (1). Associated with this is the cone K_1 of tangent directions. Associated with the functional (2) is the cone K_0 of directions of decrease. For the definitions of various terms, see Chapter 1. If H is a cone in E, then the dual cone $H^* = \{g \in E^* \mid g(\bar{p}) = 0 \forall \bar{p} \in H\}$. The Dubovitskii-Milyutin theorem states that there exist $g_0 \in K_0^*$ and $g_1 \in K_1^*$, not both zero, such that for all $(\zeta, v) \in E$, $g_0(\zeta, v) + g_1(\zeta, v) = 0$.

A. *Cone of directions of decrease.* By assumption (vii), we can minimize the alternative cost functional

$$G(\zeta, v) = \ln \int_0^T \phi_1(v,t)\,dt - \ln \int_0^T \phi_2(\zeta, v, t)\,dt. \tag{7}$$

By assumptions (iv)–(vi), the Fréchet derivative of G at (ζ_0, v_0) is given by

$$G'(\zeta_0, v_0)(\zeta, v) = \frac{\int_0^T (\nabla_v \phi_1(v_0,t), v)\,dt}{\int_0^T \phi_1(v_0,t)\,dt} - \frac{\int_0^T [(\nabla_\zeta\phi_2(\zeta_0,v_0,t),\zeta) + (\nabla_v\phi_2(\zeta_0,v_0,t),v)]\,dt}{\int_0^T \phi_2(\zeta_0,v_0,t)\,dt}. \tag{8}$$

By Theorem 7.5 of [5], (ζ, v) lies in the cone K_0 of directions of decrease if and only if $G'(\zeta_0, v_0)(\zeta, v) < 0$ or, from (8), if and only if

$$\int_0^T (\nabla_v \phi_1(v_0, t), v)\, dt - \hat{\lambda} \int_0^T (\nabla_\zeta \phi_2(\zeta_0, v_0, t), \zeta)\, dt + \hat{\lambda} \int_0^T (\nabla_v \phi_2(\zeta_0, v_0, t), v)\, dt < 0, \tag{9}$$

where $\hat{\lambda}$ is given by (5). If $K_0 \neq \emptyset$, then by Theorem 10.5 of [5], for any $g_0 \in K_0^*$,

$$g_0(\zeta, v) = -\lambda_0 \Big\{ \int_0^T (\nabla_v \phi_1(v_0, t), v)\, dt - \hat{\lambda} \int_0^T [(\nabla_\zeta \phi_2(\zeta_0, v_0, t), \zeta) + (\nabla_v \phi_2(\zeta_0, v_0, t), v)]\, dt \Big\}, \quad \lambda_0 \geq 0. \tag{10}$$

B. *Cone of tangent directions.* To find the tangent directions at (ζ_0, v_0), we apply the results of Lecture 9 of [5]. Let $\Phi(t, \tau)$ be the transition matrix corresponding to $M(t)$,

$$Q = \{(\zeta, v) \in E \mid \zeta(t) = \int_0^t \Phi(t, \tau) N(\tau) v(\tau)\, d\tau\}, \tag{11}$$

and

$$P(\zeta, v) = \zeta(t) - \int_0^t \Phi(t, \tau) N(\tau) v(\tau)\, d\tau, \tag{12}$$

which maps E into $C(0, T)$. Also the Fréchet derivative of P is

$$P'(\zeta_0, v_0)(\zeta, v) = \zeta(t) - \int_0^t \Phi(t, \tau) N(\tau) v(\tau)\, d\tau. \tag{13}$$

Clearly $P'(\zeta_0, v_0)$ is onto $C(0, T)$ (just let $v = 0$ in (13)). By Theorem 9.1 of [5], the set K_1 of tangent directions in E at (ζ_0, v_0) is given by

$$K_1 = \{(\zeta, v) \in E \mid P'(\zeta_0, v_0)(\zeta, v) = 0\}. \tag{14}$$

Thus K_1 consists of all (ζ, v) satisfying

$$\frac{d\zeta}{dt} = M(t)\zeta + N(t)v, \quad \zeta(0) = 0. \tag{15}$$

Thus K_1 is a subspace and, if $g_1 \in K_1^*$, then $g_1(\zeta, v) = 0$ for all $(\zeta, v) \in K_1$.

C. *Application of the Dubovitskii-Milyutin theorem.* Let v be arbitrary, and let ζ be a solution of (15) for this v. Then for these pairs of (ζ, v), we get (since $g_1(\zeta, v) + g_2(\zeta, v) = 0, (\zeta, v) \in E, (g_1, g_2) \not\equiv 0$)

$$\int_0^T (\nabla_v \phi_1(v_0, t), v)\, dt - \hat{\lambda} \int_0^T (\nabla_\zeta \phi_2(\zeta_0, v_0, t), \zeta)\, dt + \hat{\lambda} \int_0^T (\nabla_v \phi_2(\zeta_0, v_0, t), v)\, dt = 0 \tag{16}$$

for all v. Define ψ by

$$\frac{d\psi}{dt} = -M^*(t)\psi - \hat{\lambda}\nabla_\zeta \phi_2(\zeta_0, v_0, t), \quad \psi(T) = 0. \tag{17}$$

Then

$$\hat{\lambda} \int_0^T (\nabla_\zeta \phi_2(\zeta), v_0, t), \zeta)\, dt = -\int_0^T \left(\frac{d\psi}{dt} + M^*\psi, \zeta\right) dt = (\psi, \zeta)\,\big|_0^T + \int_0^T (\psi, Nv)\, dt. \tag{18}$$

By (16), (18), and the boundary conditions, it follows that

$$\int_0^T (\nabla_v \phi_1(v_0, t) - \hat{\lambda}\nabla_v \phi_2(\zeta_0, v_0, t) - N^*(t)\psi, v)\, dt = 0 \tag{19}$$

for any arbitrary $v \in L_p^m(0, T)$. (19) implies (6) by assumption (v).

D. *Exceptional cases.* If $K_0 = \emptyset$, we have

$$\int_0^T (\nabla_v \phi_1(v_0, t) - \hat{\lambda}\nabla_v \phi_2(\zeta_0, v_0, t) - N^*(t)\psi, v)\, dt = 0 \tag{20}$$

for all $(\zeta, v) \in E$. Thus (6) is satisfied in this case also. □

Thus we have the two point boundary value problem given by

$$\dot{\zeta} = M(t)\zeta + N(t)v, \tag{21}$$

$$\nabla_v \phi_1(v,t) - \hat{\lambda}\nabla_v \phi_2(\zeta, v, t) - N^*(t)\psi(t) = 0, \tag{22}$$

$$\dot{\psi} = -M^*(t)\psi - \hat{\lambda}\nabla_\zeta \phi_2(\zeta, v, t), \tag{23}$$

with

$$\zeta(0) = 0, \quad \psi(T) = 0. \tag{24}$$

We now give a characterization of $\hat{\lambda}$ as the minimum positive value for which the boundary value problem given by (21)–(24) has a solution with $\int_0^T \phi_2(\zeta, v, t)\, dt > 0$. The class of functionals for which this characterization is applicable will be defined in the following theorem.

THEOREM 3.2. *Let (ζ, v) be any pair that satisfies the boundary value problem given by (21)–(24) for some $\hat{\lambda}$ such that $\int_0^T \phi_2(\zeta, v, t)\, dt > 0$. In addition to assumptions (i)–(viii), assume that there exists $k > 0$ such that, for every $c \geq 0$,*

$$\begin{aligned} \phi_1(cv, t) &= c^k \phi_1(v,t), \\ \phi_2(c\zeta, cv, t) &= c^k \phi_2(\zeta, v, t). \end{aligned} \tag{25}$$

Then

$$\frac{\int_0^T \phi_1(v,t)\, dt}{\int_0^T \phi_2(\zeta, v, t)\, dt} = \hat{\lambda}. \tag{26}$$

Proof. By (25) it follows that

$$\begin{aligned} v^*\nabla_v \phi_1 &= k\phi_1(v,t), \\ v^*\nabla_v \phi_2 + \zeta^*\nabla_\zeta \phi_2 &= k\phi_2(\zeta, v, t). \end{aligned} \tag{27}$$

From (23),

$$\int_0^T \zeta^*\dot{\psi}\, dt = -\int_0^T \zeta^* M^* \psi\, dt - \hat{\lambda}\int_0^T \zeta^*\nabla_\zeta \phi_2\, dt. \tag{28}$$

Integrating the left side of (28) by parts and making use of (21) and (24), we get

$$\int_0^T \zeta^* \dot{\psi}\, dt = -\int_0^T \zeta^* M^* \psi\, dt - \int_0^T v^* N^* \psi\, dt. \tag{29}$$

From (22),

$$\int_0^T v^* N^* \psi\, dt = \int_0^T v^* \nabla_v \phi_1\, dt - \hat{\lambda} \int_0^T v^* \nabla_v \phi_2\, dt. \tag{30}$$

From (28)–(30), we get

$$\int_0^T v^* \nabla_v \phi_1\, dt = \hat{\lambda} \left[\int_0^T v^* \nabla_v \phi_2\, dt + \int_0^T \zeta^* \nabla_\zeta \phi_2\, dt \right]. \tag{31}$$

From (27) and (31), we get

$$\frac{\int_0^T \phi_1(v,t)\, dt}{\int_0^T \phi_2(\zeta, v, t)\, dt} = \hat{\lambda}.$$

Thus the proof is complete. □

Since the minimizing $v(t)$ satisfies (21)–(24), we deduce from Theorem 3.2 that the minimum value of the functional in (2) is the same as the minimum positive value of $\hat{\lambda}$ for which (21)–(24) has a solution with $\int_0^T \phi_2(\zeta, v, t)\, dt > 0$.

4. Variation of the Minimum Value

Many practical systems undergo variations in parameters. In such circumstances, it is important to know the variation of $\hat{\lambda}$ in terms of the parameter variations so that we have an idea of the degradation in performance of the system. In this section we derive such an expression. In the following section the application of this expression to the H_∞ performance robustness problem will be given.

Consider (1) and (2). In addition to assumptions (i)–(vii) of Section 3, we make the following assumptions.

(a) $\nabla_{vv}\phi_1, \nabla_{vv}\phi_2, \nabla_{\zeta\zeta}\phi_2, \nabla_{\zeta v}\phi_2$ exist and are continuous in (ζ, v), and all are measurable in t.

(b) There exists $k > 0$ such that for all $c \geq 0$,

$$\begin{aligned} \phi_1(cv, t) &= c^k \phi_1(v, t), \\ \phi_2(c\zeta, cv, t) &= c^k \phi_2(\zeta, v, t). \end{aligned} \tag{32}$$

Note that $\hat{\lambda} = \inf_{v \neq 0} F(\zeta, v)$. Let the elemental variations in M and N be denoted by δM and δN, respectively. Let $\delta\hat{\lambda}$ be the variation in $\hat{\lambda}$ due to δM and δN. We now derive an expression for $\delta\hat{\lambda}$ in terms of $\delta M(t)$ and $\delta N(t)$. For simplicity of notation, we omit displaying the arguments of the functions in the following analysis.

Let v minimize (2). From Section 3, we get the following two point boundary value problem which needs to be satisfied by $\zeta(t)$ and the adjoint vector $\psi(t)$.

$$\dot{\zeta} = M\zeta + Nv \tag{33}$$

$$\nabla_v \phi_1 - \hat{\lambda} \nabla_v \phi_2 - N^* \psi = 0 \tag{34}$$

$$\dot{\psi} = -M^* \psi - \hat{\lambda} \nabla_\zeta \phi_2 \tag{35}$$

$$\zeta(0) = 0, \qquad \psi(T) = 0. \tag{36}$$

Let ζ_1 and ψ_1 respectively represent the variations in ζ and ψ due to δM and δN. Also, let v_1 be the corresponding variation in v. From (33)–(36), we have the following equations that are satisfied by ζ_1, ψ_1, and v_1.

$$\dot{\zeta}_1 = M\zeta_1 + \delta M \zeta + N v_1 + \delta N v, \tag{37}$$

$$(\nabla_{vv}\phi_1 - \hat{\lambda}\nabla_{vv}\phi_2) v_1 - \hat{\lambda}\nabla_{v\zeta}\phi_2 \zeta_1 - N^*\psi_1 - \delta\hat{\lambda}\nabla_v\phi_2 - \delta N^* \psi = 0, \tag{38}$$

$$\dot{\psi}_1 = -M^*\psi_1 - \delta M^* \psi - \hat{\lambda}\nabla_{\zeta\zeta}\phi_2 \zeta_1 - \hat{\lambda}\nabla_{\zeta v}\phi_2 v_1 - \delta\hat{\lambda}\nabla_\zeta \phi_2, \tag{39}$$

$$\zeta_1(0) = 0, \qquad \psi_1(T) = 0. \tag{40}$$

THEOREM 4.1. *Consider (33)–(40). Then the variation in $\hat{\lambda}$ is given by*

$$\delta\hat{\lambda} = \frac{-\int_0^T (\delta M\,\zeta + \delta N\,v, \psi)\,dt}{\int_0^T \phi_2\,dt}. \tag{41}$$

Proof. From (39), we get

$$\int_0^T (\zeta, \dot{\psi}_1)\,dt = -\int_0^T \left[(\zeta, M^*\psi_1) - (\zeta, \delta M^*\psi) - \hat{\lambda}(\zeta, \nabla_{\zeta\zeta}\phi_2\zeta_1)\right] dt$$
$$-\hat{\lambda}\int_0^T (\zeta, \nabla_{\zeta v}\phi_2 v_1)\,dt - \delta\hat{\lambda}\int_0^T (\zeta, \nabla_\zeta\phi_2)\,dt. \tag{42}$$

Also, by an integration by parts and by (33), (36), and (40),

$$\int_0^T (\zeta, \dot{\psi}_1)\,dt = -\int_0^T (\zeta, M^*\psi_1)\,dt - \int_0^T (v, N^*\psi_1)\,dt. \tag{43}$$

From (42) and (43), we get

$$\delta\hat{\lambda}\int_0^T (\zeta, \nabla_\zeta\phi_2)\,dt + \int_0^T (\zeta, \delta M^*\psi)\,dt + \hat{\lambda}\int_0^T (\zeta, \nabla_{\zeta\zeta}\phi_2\zeta_1)\,dt$$
$$+\hat{\lambda}\int_0^T (\zeta, \nabla_{\zeta v}\phi_2 v_1)\,dt = \int_0^T (v, N^*\psi_1)\,dt. \tag{44}$$

From (38), we get

$$\int_0^T (v, N^*\psi_1)\,dt = \int_0^T (v, (\nabla_{vv}\phi_1 - \hat{\lambda}\nabla_{vv}\phi_2)v_1 - \hat{\lambda}\nabla_{v\zeta}\phi_2\zeta_1)\,dt$$
$$-\int_0^T (v, \delta N^*\psi)\,dt - \delta\hat{\lambda}\int_0^T (v, \nabla_v\phi_2)\,dt. \tag{45}$$

Substituting (45) in the right side of (44) and making use of the relations

$$(v, \nabla_v\phi_2) + (\zeta, \nabla_\zeta\phi_2) = k\phi_2,$$

$$(v, \nabla_{vv}\phi_1 v_1) = (k-1)(\nabla_v\phi_1, v_1),$$

$$(\zeta, \nabla_{\zeta\zeta}\phi_2\zeta_1) + (v, \nabla_{v\zeta}\phi_2\zeta_1) = (k-1)(\nabla_\zeta\phi_2, \zeta_1),$$

$$(\zeta, \nabla_{\zeta v}\phi_2 v_1) + (v, \nabla_{vv}\phi_2 v_1) = (k-1)(\nabla_v\phi_2, v_1),$$

we get

$$k\delta\hat{\lambda}\int_0^T \phi_2\,dt + \int_0^T (\delta M\,\zeta + \delta N\,v, \psi)\,dt + \hat{\lambda}(k-1)\int_0^T (\nabla_\zeta \phi_2, \zeta_1)\,dt$$
$$= (k-1)\int_0^T (\nabla_v \phi_1 - \hat{\lambda}\nabla_v \phi_2, v_1)\,dt. \quad (46)$$

Note that from (34),

$$\nabla_v \phi_1 - \hat{\lambda}\nabla_v \phi_2 = N^*\psi. \quad (47)$$

Also by (35),

$$\hat{\lambda}(k-1)\int_0^T (\nabla_\zeta \phi_2, \zeta_1)\,dt = -(k-1)\int_0^T (\dot{\psi} + M^*\psi, \zeta_1)\,dt. \quad (48)$$

By an integration by parts, (48) becomes

$$\hat{\lambda}(k-1)\int_0^T (\nabla_\zeta \phi_2, \zeta_1)\,dt = (k-1)\int_0^T (\psi, \delta M\,\zeta + N\,v_1 + \delta N\,v)\,dt. \quad (49)$$

Substituting (47) and (49) in (46), we get

$$\delta\hat{\lambda}\int_0^T \phi_2\,dt + \int_0^T (\delta M\,\zeta + \delta N\,v, \psi)\,dt = 0. \quad (50)$$

Thus the proof is complete. □

5. Application to Performance Robustness

We will use the previous theory to generate an expression for performance variation in terms of parameter variations. This expression will be useful for the approximate solution of certain finite horizon H_∞ performance robustness problems. The idea is to make sure that the performance does not fall below the required level, even under variations in the system matrices.

Expressions for a suboptimal H_∞ controller are given in [6] and [7] under various assumptions. We now give expressions for a suboptimal finite horizon

H_∞ controller in a general setting. These expressions were derived in Chapter 3, which is based on [8].

The system is given by

$$\dot{x} = A(t)x + B_1(t)u + B_2(t)v, \quad x(0) = 0, \tag{51}$$

$$z = C(t)x + D(t)u + E(t)v, \tag{52}$$

$$y = C_2(t)x + D_2(t)u + E_2(t)v. \tag{53}$$

Given $\lambda > 0$, the finite horizon H_∞ suboptimal control problem is to select a controller such that

$$\min_{v \neq 0} \frac{\int_0^T \frac{1}{2} v^* R v \, dt}{\int_0^T \frac{1}{2} z^* W z \, dt} > \lambda. \tag{54}$$

Let $W_1, \ldots, W_6$ be defined by

$$z^* W z = x^* W_1 x + 2x^* W_2 u + u^* W_3 u + 2x^* W_4 v + v^* W_5 v + 2u^* W_6 v. \tag{55}$$

Let

$$\tilde{z}_1 = B_2^* \tilde{e} + E_2^* \tilde{u} + E^* \tilde{w}, \tag{56}$$

and let $\tilde{W}_1, \ldots, \tilde{W}_6$ be defined by

$$\tilde{z}_1^* R^{-1} \tilde{z}_1 = \tilde{e}^* \tilde{W}_1 \tilde{e} + 2\tilde{e}^* \tilde{W}_2 \tilde{u} + \tilde{u}^* \tilde{W}_3 \tilde{u} + 2\tilde{e}^* \tilde{W}_4 \tilde{w} + \tilde{w}^* \tilde{W}_5 \tilde{w} + 2\tilde{u}^* \tilde{W}_6 \tilde{w}. \tag{57}$$

We make the following assumptions.

(a) Assume that for each $t \in [0, T]$, $R - \lambda W_5$ is positive definite and $W_3 + \lambda W_6 (R - \lambda W_5)^{-1} W_6^*$ is invertible.

(b) Also assume that for each $t \in [0, T]$, $W^{-1} - \lambda W_5$ is positive definite and $\tilde{W}_3 + \lambda \tilde{W}_6 (W^{-1} - \lambda \tilde{W}_5)^{-1} \tilde{W}_6^*$ is invertible.

The relevant controller equations are

$$\Omega = (R - \lambda W_5)^{-1}, \tag{58}$$

$$\Lambda = (W_3 + \lambda W_6 \Omega W_6^*)^{-1}, \tag{59}$$

$$U_1 = \Lambda(B_1^* + \lambda W_6 \Omega B_2^*), \tag{60}$$

$$U_2 = -\Lambda(W_2^* + \lambda W_6 \Omega W_4^*), \tag{61}$$

$$V_1 = \lambda\Omega(-B_2^* + W_6^* U_1), \tag{62}$$

$$V_2 = \lambda\Omega(W_4^* + W_6^* U_2), \tag{63}$$

$$\dot{P} + P(A + B_1U_2 + B_2V_2) + (A^* - W_2U_1 - W_4V_1)P$$
$$+P(B_1U_1 + B_2V_1)P - (W_1 + W_2U_2 + W_4V_2) = 0, \quad P(T) = 0. \tag{64}$$

Note that the above equation is symmetric.

The relevant observer equations are

$$\Phi = (W^{-1} - \lambda\tilde{W}_5)^{-1}, \tag{65}$$

$$\Gamma = (\tilde{W}_3 + \lambda\tilde{W}_6\Phi\tilde{W}_6^*)^{-1}, \tag{66}$$

$$\tilde{A} = A + B_2V_1P + B_2V_2, \tag{67}$$

$$\tilde{B} = -(U_1P + U_2), \tag{68}$$

$$\tilde{C} = C_2 + E_2V_1P + E_2V_2, \tag{69}$$

$$S_1 = \Gamma(\tilde{C} + \lambda\tilde{W}_6\Phi D\tilde{B}), \tag{70}$$

$$S_2 = -\Gamma(\tilde{W}_2^* + \lambda\tilde{W}_6\Phi\tilde{W}_4^*), \tag{71}$$

$$T_1 = \lambda\Phi(-D\tilde{B} + \tilde{W}_6^*S_1), \tag{72}$$

$$T_2 = \lambda\Phi(\tilde{W}_4^* + \tilde{W}_6^*S_2), \tag{73}$$

$$\dot{Y} = (\tilde{A} - \tilde{W}_2S_1 - \tilde{W}_4T_1)Y + Y(\tilde{A}^* + \tilde{C}^*S_2 + \tilde{B}^*D^*T_2)$$
$$+Y(\tilde{C}^*S_1 + \tilde{B}^*D^*T_1)Y - (\tilde{W}_1 + \tilde{W}_2S_2 + \tilde{W}_4T_2), \quad Y(0) = 0. \tag{74}$$

Note that the above equation is symmetric.

A suboptimal controller is given by

$$\dot{q} = Aq + B_1(U_1Pq + U_2q) + B_2(V_1Pq + V_2q) + L(\tilde{C}q + D_2u - y), \tag{75}$$

$$L = (S_1Y + S_2)^*, \tag{76}$$

$$u = (U_1P + U_2)q. \tag{77}$$

In the case of time-invariant systems, the solutions of the Riccati equations (64) and (74) eventually converge to constant matrices. Note that the full state feedback controller is simply given by $u = (U_1P + U_2)x$.

From the above expressions, the closed loop system is given by

$$\begin{pmatrix} \dot{x} \\ \dot{q} \end{pmatrix} = \begin{pmatrix} A & B_1(U_1P + U_2) \\ -LC_2 & \tilde{U} - LD_2(U_1P + U_2) \end{pmatrix} \begin{pmatrix} x \\ q \end{pmatrix} + \begin{pmatrix} B_2 \\ -LE_2 \end{pmatrix} v, \tag{78}$$

where

$$\tilde{U} = A + B_1(U_1P + U_2) + B_2(V_1P + V_2) + L(\tilde{C} + D_2(U_1P + U_2)). \tag{79}$$

We now specialize to the case where

$$x(0) = q(0) = 0. \tag{80}$$

For the controller given by (77), we have

$$z = Cx + D(U_1P + U_2)q + Ev. \tag{81}$$

The performance of the controller is given by

$$\min_{v \neq 0} \frac{\int_0^T \frac{1}{2} v^* R v \, dt}{\int_0^T \frac{1}{2} z^* W z \, dt}, \tag{82}$$

which is strictly greater than λ by our design procedure. In the time-invariant case, the solutions of (64) and (74) basically are the solutions of the corresponding algebraic Riccati equations.

Let (78) be written as

$$\dot{\zeta} = M(t)\zeta + N(t)v, \quad \zeta(0) = 0. \tag{83}$$

Assume that in (52), $C^*D = 0$ and $C^*E = 0$, so that

$$z^*Wz = \zeta^*Q_1\zeta + v^*Q_2v. \tag{84}$$

The exogenous input v needs to be chosen to minimize

$$J(v) = \frac{\int_0^T \frac{1}{2}v^*Rv\,dt}{\int_0^T \frac{1}{2}\{\zeta^*Q_1\zeta + v^*Q_2v\}\,dt}. \tag{85}$$

The results of the earlier sections are now applicable to this problem. Let $\hat{\lambda} = \min_{v\neq 0} J(v)$. Let v minimize (85) and ψ denote the adjoint vector. From (21)–(24), we get

$$\dot{\zeta} = M(t)\zeta + N(t)v, \tag{86}$$

$$v = \left(R(t) - \hat{\lambda}Q_2(t)\right)^{-1}N^*(t)\psi, \tag{87}$$

$$\dot{\psi} = -M^*(t)\psi - \hat{\lambda}Q_1(t)\zeta, \tag{88}$$

$$\zeta(0) = 0, \qquad \psi(T) = 0. \tag{89}$$

From Theorem 3.2, we can deduce that $\hat{\lambda}$ is the smallest positive value for which the boundary value problem

$$\begin{pmatrix} \dot{\zeta} \\ \dot{\psi} \end{pmatrix} = \begin{pmatrix} M & N(R - \hat{\lambda}Q_2)^{-1}N^* \\ -\hat{\lambda}Q_1 & -M^* \end{pmatrix} \begin{pmatrix} \zeta \\ \psi \end{pmatrix}, \tag{90}$$

$$\zeta(0) = 0, \qquad \psi(T) = 0. \tag{91}$$

has a solution with $\int_0^T \{\zeta^*Q_1(t)\zeta + v^*Q_2(t)v\}\,dt > 0$. This criterion can be used to evaluate $\hat{\lambda}$.

Let $\delta\hat{\lambda}$ be the variation in $\hat{\lambda}$ due to δM and δN. From Theorem 4.1, $\delta\hat{\lambda}$ is given by

$$\delta\hat{\lambda} = \frac{-\int_0^T (\delta M(t)\zeta + \delta N(t)v, \psi)\, dt}{\int_0^T \frac{1}{2}\{\zeta^* Q_1(t)\zeta + v^* Q_2(t) v\}\, dt}. \tag{92}$$

A similar expression is derived in Chapter 4, which is based on [9].

Using the theory above, an iterative procedure similar to the one in Chapter 4 can be given to approximately solve the suboptimal performance robustness problem.

6. Conclusions

A general minimization problem is treated in which the cost functional is a quotient of definite integrals. An existence result and necessary conditions for the minimizer are given. The results are useful for the computation of the H_∞ norm of the closed loop system for a given controller. Also, an expression for the variation of performance is derived in terms of variations in the system matrices. The results can be applied for the approximate solution of the finite horizon H_∞ performance robustness problem.

References

[1] Subrahmanyam, M. B., "On applications of control theory to integral inequalities," *Journal of Mathematical Analysis and Applications*, Vol. 77, 1980, pp. 47–59.

[2] Subrahmanyam, M. B., "On applications of control theory to integral inequalities: II," *SIAM Journal on Control and Optimization*, Vol. 19, 1981, pp. 479–489.

[3] SUBRAHMANYAM, M. B., "On integral inequalities associated with a linear operator equation," *Proceedings of the American Mathematical Society*, Vol. 92, 1984, pp. 342–346.

[4] LEE, E. B. AND MARKUS, L., *Foundations of Optimal Control Theory*, John Wiley, New York, 1967.

[5] GIRSANOV, I. V., *Lectures on Mathematical Theory of Extremum Problems*, Lecture Notes in Economics and Mathematical Systems, No. 67, Springer-Verlag, Berlin, 1972.

[6] DOYLE, J. C., GLOVER, K., KHARGONEKAR, P. P., AND FRANCIS, B. A., "State-space solutions to standard H_2 and H_∞ control problems," *IEEE Transactions on Automatic Control*, Vol. 34, 1989, pp. 831–847.

[7] RAVI, R., NAGPAL, K. M. AND KHARGONEKAR, P. P., "H_∞ control of linear time-varying systems: a state space approach," *SIAM Journal on Control and Optimization*, Vol. 29, 1991, pp. 1394–1413.

[8] SUBRAHMANYAM, M. B., "General formulae for suboptimal H_∞ control over a finite horizon," *International Journal of Control*, Vol. 57, No. 2, 1993, pp. 365–375.

[9] SUBRAHMANYAM, M. B., "Finite Horizon H_∞ with parameter variations," *International Journal of Robust and Nonlinear Control*, Vol. 4, No. 5, September–October 1994, pp. 631–643.

Chapter 6

H_∞ Design of the F/A-18A Automatic Carrier Landing System

Abstract

In this chapter a design of the F/A-18A Automatic Carrier Landing System is accomplished using finite horizon H_∞ techniques. If the final time is sufficiently large, the dynamic Riccati equations involved in the design of the suboptimal output feedback controller give rise to constant solutions. Only longitudinal equations of motion are considered, and thrust is incorporated as a control variable. The object of the design is to maintain a constant flight path angle under worst-case conditions of vertical gust and sensor noise during carrier landing. The design yields satisfactory response for vertical rate command as well.

1. Introduction

The Navy Automatic Carrier Landing System (ACLS) incorporates a radar and a computer on board the ship, and a data link to the aircraft autopilot. The computer computes pitch and bank commands from the airplane space position and ship motion data, and transmits them to the aircraft via a radio frequency data link.

In order to keep the touchdown error small, the H-dot control law was devised [1-3]. The H-dot control law was also observed to significantly reduce flight path errors due to ship burble. The aircraft also needs to follow the vertical motion of touchdown point on the ship. Deck Motion Compensation (DMC) is used to filter the calculated deck position. The Approach Power Compensation System (APCS) on an F/A-18A contains feedback gains from normal accelerometer, pitch rate, stabilizer position, and bank angle, in addition to the angle of attack error. The primary feedback signal for the APCS is the angle of attack error. This error drives the servo and thus modulates the engine thrust to maintain the angle of attack during an approach.

The aircraft needs to arrive at the touchdown point with proper sink speed and position in space to closely match the position and vertical motion of the carrier deck touchdown zone. Aircraft hook should impact the deck between No. 2 and No. 3 arresting cables. The sink speed must be 10–14 ft/sec. In this chapter the flight condition is at 136 kts with a flight path angle of $-3°$, giving rise to a sink speed of 12 ft/sec. Also, the trim angle of attack α_{trim} is 8.3°.

In the H_∞ suboptimal design proposed here, we make use of the results of Chapter 3, which is based essentially on [4], to synthesize an output feedback controller. In the case of output feedback, the suboptimal value that gives a viable design is considerably greater than the infimal H_∞ norm with state feedback. Also, in the case of output feedback, considerable experimentation with the noninfimal value and the weighting matrices was required to obtain a satisfactory design.

The objectives of the design are to keep the angle of attack and the flight path angle at the reference values. The disturbance terms are vertical gust, carrier air wake, and sensor noise. The main aim of the control law design is to minimize the flight path error due to turbulence.

2. H_∞ Controller Design

The longitudinal small perturbation equations of F/A-18A at 136 kts and an altitude of 50 ft with full flaps are given by

$$\dot{x} = Ax + B_1u + B_2v, \tag{1}$$

where the system matrices A, B_1, and B_2 are as given in Table 1 [3]. In (1),

$$x = (\,\bar{u}/V \quad \alpha \quad \theta \quad q \quad h/V\,)^*,$$

$$u = (\,\delta_H \quad \delta_{LEF} \quad \delta_{RT} \quad \delta_{PL}\,)^*,$$

and

$$v = (\,\alpha_g \quad v_1 \quad v_2 \quad v_3 \quad v_4\,)^*,$$

where $\bar{u}/V$ is the perturbed normalized forward velocity, α is the perturbed angle of attack in radians, θ is the perturbed pitch angle in radians, q is the perturbed pitch rate in radians/sec, h/V is the perturbed normalized altitude in seconds, δ_H is the perturbed horizontal tail deflection in radians, δ_{LEF} is the perturbed leading edge flap deflection in radians, δ_{RT} is the perturbed rudder toe-in deflection in radians, δ_{PL} is the engine power lever control angle in degrees, α_g is the incremental angle of attack due to vertical gust in radians, and v_1, v_2, v_3, v_4 are sensor noises.

The output and error equations are given by

$$y = C_2x + D_2u + E_2v, \tag{2}$$

$$z = Cx + Du + Ev, \tag{3}$$

where the matrices C_2, D_2, E_2, C, D, and E are as given in Table 2. In this chapter the superscript $*$ denotes a matrix or a vector transpose. The square of the infimal H_∞ norm is given by the inverse of

$$\min_{v \neq 0} \max_{u} \frac{\int_0^{25} v^* R v \, dt}{\int_0^{25} z^* W z \, dt},$$

where $R = 1000 I_5$ and $W = I_6$, I_5 and I_6 being identity matrices. Denote the value of the above expression by λ_{opt}.

Table 1 Plant matrices of the F/A-18A longitudinal system

$$A = \begin{pmatrix} -0.0705 & 0.0475 & -0.1403 & 0.0000 & -0.000058 \\ -0.3110 & -0.3430 & 0.0000 & 0.99133 & 0.00102 \\ 0.0000 & 0.0000 & 0.0000 & 1.0000 & 0.0000 \\ 0.0218 & -1.1660 & 0.0000 & -0.2544 & 0.0000 \\ 0.0000 & -1.0000 & 1.0000 & 0.0000 & 0.0000 \end{pmatrix}$$

$$B_1 = \begin{pmatrix} 0.0121 & 0.00248 & 0.1690 & 0.2316 \\ -0.0721 & 0.0140 & 0.0128 & -0.0338 \\ 0.0000 & 0.0000 & 0.0000 & 0.0000 \\ -1.8150 & -0.0790 & 0.1681 & 0.0023 \\ 0.0000 & 0.0000 & 0.0000 & 0.0000 \end{pmatrix}$$

$$B_2 = \begin{pmatrix} 0.0475 & 0 & 0 & 0 & 0 \\ -0.343 & 0 & 0 & 0 & 0 \\ 0 & 0 & 0 & 0 & 0 \\ -1.166 & 0 & 0 & 0 & 0 \\ 0 & 0 & 0 & 0 & 0 \end{pmatrix}$$

Table 2 Matrices associated with the output and the error

$$C_2 = \begin{pmatrix} 0 & 0 & 0 & 0 & 1 \\ 0 & -1 & 1 & 0 & 0 \\ 0.311 & 0.343 & 0 & 0.0087 & -0.001 \\ 0 & 1 & 0 & 0 & 0 \end{pmatrix}$$

$$D_2 = \begin{pmatrix} 0 & 0 & 0 & 0 \\ 0 & 0 & 0 & 0 \\ 0.0721 & -0.014 & -0.0128 & 0.0338 \\ 0 & 0 & 0 & 0 \end{pmatrix}$$

$$E_2 = \begin{pmatrix} 0 & 1 & 0 & 0 & 0 \\ 0 & 0 & 1 & 0 & 0 \\ 0.343 & 0 & 0 & 1 & 0 \\ 0 & 0 & 0 & 0 & 1 \end{pmatrix}, \quad C = \begin{pmatrix} -5 & 17 & 0 & 0 & 0 \\ 0 & 0 & 0 & 0 & 35 \\ 0 & 0 & 0 & 0 & 0 \\ 0 & 0 & 0 & 0 & 0 \\ 0 & 0 & 0 & 0 & 0 \\ 0 & 0 & 0 & 0 & 0 \end{pmatrix}$$

$$D = \begin{pmatrix} 0 & 0 & 0 & 0 \\ 0 & 0 & 0 & 0 \\ 0.075 & 0 & 0 & 0 \\ 0 & 0.5 & 0 & 0 \\ 0 & 0 & 0.5 & 0 \\ 0 & 0 & 0 & 0.4 \end{pmatrix}, \quad E = \begin{pmatrix} 0 & 0 & 0 & 0 & 0 \\ 0 & 0 & 0 & 0 & 0 \\ 0 & 0 & 0 & 0 & 0 \\ 0 & 0 & 0 & 0 & 0 \\ 0 & 0 & 0 & 0 & 0 \\ 0 & 0 & 0 & 0 & 0 \end{pmatrix}$$

The suboptimal controller synthesis problem can be stated as follows. Given $\lambda < \lambda_{\text{opt}}$, find an output feedback controller, if one exists, for which

$$\min_{v \neq 0} \frac{\int_0^{25} v^* R v \, dt}{\int_0^{25} z^* W z \, dt} > \lambda. \tag{4}$$

Thus, the suboptimal controller keeps the norm of the transfer function from the sensor and turbulence disturbances to the error z within a specified value. The design assures that the error energy will be bounded by a specified constant multiplied by the energy of the disturbance terms.

For the output feedback controller, the variables utilized for feedback are h/V, $\dot{h}/V$, $\ddot{h}/V$, and α. The general aims are to make the system sufficiently stable and to keep the steady state perturbed angle of attack near zero under a step vertical rate command. Considerable experimentation with the value of λ and the matrices C and D was needed to obtain a satisfactory design. With $\lambda = 0.5$, the solutions of the two dynamic Riccati equations given by (107) and (117) of Chapter 3 eventually became constant. The gain of the filter L in (118) of Chapter 3 is

$$L = \begin{pmatrix} 0.0141 & -0.0118 & -0.0113 & 0.0419 \\ 0.2036 & 0.1385 & -0.0818 & -0.6514 \\ -0.0875 & -0.0095 & -0.1513 & -0.5129 \\ 0.1139 & 0.2223 & 0.2407 & -0.3456 \\ -1.4786 & -0.2911 & 0.0449 & 0.2036 \end{pmatrix}. \tag{5}$$

The control feedback matrix $U_1P + U_2$ in (120) of Chapter 3 is given by

$$U_1P + U_2 = \begin{pmatrix} -57.8770 & 132.2905 & 94.5214 & 12.3637 & 95.8467 \\ 1.0558 & -12.6624 & 13.2678 & 0.5010 & 21.6190 \\ -2.2725 & 6.4705 & -5.0668 & -0.1977 & 0.6566 \\ -9.4966 & 64.1184 & -61.5367 & -2.2647 & -81.3981 \end{pmatrix}. \tag{6}$$

The output feedback control law is given by $u = (U_1P + U_2)q$. Let

$$\tilde{U} = A + B_1(U_1P + U_2) + B_2(V_1P + V_2) + L(\tilde{C} + D_2(U_1P + U_2)).$$

The closed loop output feedback system is given by

$$\begin{pmatrix} \dot{x} \\ \dot{q} \end{pmatrix} = \begin{pmatrix} A & B_1(U_1P+U_2) \\ -LC_2 & \tilde{U} - LD_2(U_1P+U_2) \end{pmatrix} \begin{pmatrix} x \\ q \end{pmatrix} + \begin{pmatrix} B_2 \\ -LE_2 \end{pmatrix} v. \quad (7)$$

For the definition of the various matrices in the above equation, see Chapter 3. Let

$$A_c = \begin{pmatrix} A & B_1(U_1P+U_2) \\ -LC_2 & \tilde{U} - LD_2(U_1P+U_2) \end{pmatrix}.$$

The eigenvalues of the closed loop system matrix A_c are given by $-16.8834 \pm i11.3506$, $-1.3228 \pm i1.8574$, $-0.7011 \pm i1.0847$, $-0.1391 \pm i0.2222$, -1.2911, and -1.8509.

For vertical rate response to a step command of 5 ft/sec, the equation is given by

$$\begin{pmatrix} \dot{x} \\ \dot{q} \end{pmatrix} = A_c \begin{pmatrix} x \\ q \end{pmatrix} - r,$$

where $r = (0 \;\; 0 \;\; 0 \;\; 0 \;\; 0.0218 \;\; 0 \;\; 0 \;\; 0 \;\; 0 \;\; 0.0218)^*$. The vertical rate response was observed to get close to the steady state value in about 4 seconds.

3. Actuator and Engine Dynamics

The actuator dynamics are fast enough that they do not significantly alter the effect of the controller. That is, when we augmented the actuator dynamics to the system dynamics given in the previous section, and applied the lower-order output feedback controller synthesized above, there was not a significant change in the responses. However, the engine dynamics are slow enough that we need to verify the adequacy of the controller with the engine dynamics included.

The low-order approximations for the actuator and engine models are given in the following page. The subscript c refers to the command. Also, δ_T refers to the equivalent thrust angle.

Stabilator:

$$\frac{\delta_H}{\delta_{Hc}} = \frac{1325}{s^2 + 29.85s + 1325}$$

Leading edge flap:

$$\frac{\delta_{LEF}}{\delta_{LEFc}} = \frac{2230}{s^2 + 109.8s + 2230}$$

Rudder:

$$\frac{\delta_{RT}}{\delta_{RTc}} = \frac{(72.1)^2}{s^2 + 2(0.69)(72.1)s + (72.1)^2}$$

APCS servo:

$$\frac{\delta_{PL}}{\delta_{PLc}} = \frac{1100}{s^2 + 33.17s + 1100}$$

Engine:

$$\frac{\delta_T}{\delta_{PL}} = \frac{2.994(s^3 + 3.5s^2 + 9.18s + 3.13)}{s^4 + 6.5s^3 + 18.25s^2 + 26.28s + 9.37}$$

Note that in (1), δ_{PL} needs to be actually replaced by δ_T in case the engine dynamics are to be included.

The state-space representations are given below.

Stabilator:

$$\dot{x}_1 = x_2$$

$$\dot{x}_2 = -29.85x_2 - 1325x_1 + 1325\delta_{hc}$$

Leading edge flap:

$$\dot{x}_3 = x_4$$

$$\dot{x}_4 = -109.8x_4 - 2230x_3 + 2230\delta_{LEFc}$$

Rudder:

$$\dot{x}_5 = x_6$$

$$\dot{x}_6 = -99.5x_6 - 5200x_5 + 5200\delta_{RTc}$$

APCS:

$$\dot{x}_7 = x_8 + x_{11} - 6.5x_7$$

$$\dot{x}_8 = x_9 + 3.5x_{11} - 18.25x_7$$

$$\dot{x}_9 = x_{10} + 9.18x_{11} - 26.38x_7$$

$$\dot{x}_{10} = 3.13x_{11} - 9.37x_7$$

$$\dot{x}_{11} = x_{12}$$

$$\dot{x}_{12} = -33.17x_{12} - 1100x_{11} + 3293\delta_{PL_c}$$

We augment (1) with the above state space dynamics, letting $\delta_H = x_1$, $\delta_{LEF} = x_3$, $\delta_{RT} = x_5$, and $\delta_{PL} = x_7$ in (1). The augmented 17-dimensional system is denoted by

$$\dot{x}_s = A_s x_s + B_{1_s} u_s + B_{2_s} v, \tag{8}$$

where

$$u_s = (\,\delta_{H_c} \quad \delta_{LEF_c} \quad \delta_{RT_c} \quad \delta_{PL_c}\,). \tag{9}$$

Let Z represent the 4×12 matrix consisting of zero elements. With the output feedback controller of the previous section, the closed loop system is given by

$$\begin{pmatrix} \dot{x}_s \\ \dot{q} \end{pmatrix} = A_{\text{sys}} \begin{pmatrix} x_s \\ q \end{pmatrix},$$

where

$$A_{\text{sys}} = \begin{pmatrix} A_s & B_{1_s}(U_1 P + U_2) \\ -L\,(\,C_2 \quad Z\,) & \tilde{U} - LD_2(U_1 P + U_2) \end{pmatrix}. \tag{10}$$

The eigenvalues of A_{sys} are given by -82.9, $-49.7 \pm i52.2$, $-15 \pm i33$, $-16.6 \pm i28.7$, -26.9, $-16.8 \pm i11.4$, $-3.2 \pm i0.8$, $-1.7 \pm i2.2$, $-0.53 \pm i1.84$, $-0.52 \pm i0.92$, -1.66, $-0.15 \pm i0.31$, -0.21. With the lower-order controller given by $u = (U_1 P + U_2)q$, the vertical rate command response of the augmented full model approaches steady-state value in approximately 6 seconds. The responses are shown in Figs. 1 and 2. Alternately, the H_∞ output feedback design can be carried out using the augmented full model. However, in this case the controller will be of higher order also. The perturbed deflections of the stabilator, leading edge flap, rudder toe-in, and the engine power lever control angle are shown in Figs. 3–6.

It is desirable in the case of an ACLS that the aircraft respond quickly to vertical rate commands, especially as the aircraft approaches close to the carrier. From Fig. 1, it can be observed that the response is fast, although for a short duration it is in the wrong direction because of the presence of a nonminimum phase zero in the transfer function from the stabilator to the vertical rate. Under the step vertical rate command of 5 ft/sec, the H_∞ design overshoots to 6 ft/sec, in contrast to the existing H-dot design which overshoots to 6.5 ft/sec [2].

Figs. 7 and 8 give the frequency response plots of vertical rate to vertical rate command. The responses meet AR-40A specifications [5].

4. Response to Disturbances

We now evaluate the response of the output feedback law

$$u = (U_1 P + U_2) q \tag{11}$$

with $U_1 P + U_2$ given by (6), to turbulence and vertical burble. We ignore in this section the actuator and engine dynamics and consider the closed loop output feedback system given by

$$\begin{pmatrix} \dot{x} \\ \dot{q} \end{pmatrix} = A_c \begin{pmatrix} x \\ q \end{pmatrix} + B_c \alpha_g, \tag{12}$$

where A_c is defined in Section 2, and B_c is the first column of the matrix $\begin{pmatrix} B_2 \\ -LE_2 \end{pmatrix}$, and α_g is the incremental angle of attack due to vertical gust.

The transfer function $G(s)$ from α_g to the vertical displacement h can be obtained from (12). The power spectral density of h is

$$\Phi_h = |G(j\omega)|^2 \Phi_{\alpha_g},$$

where Φ_{α_g} is taken from the ACLS system specification AR-40A [5] to be

$$\Phi_{\alpha_g} = \frac{71.6}{V_0^3 \left[1 + \left(\frac{100\omega}{V_0}\right)^2\right]}.$$

The variance of vertical error at touchdown due to vertical turbulence is

$$\sigma_h^2 = \int_{-\infty}^{\infty} \Phi_h(\omega)\, d\omega.$$

The value of σ_h is computed to be 0.19 ft. For a 3° glide slope, the ratio of longitudinal velocity to sink rate is approximately 19 to 1. Thus the longitudinal touchdown dispersion is approximately 3.6 ft, which is quite satisfactory. The longitudinal touchdown dispersion with the H-dot control law is around 8 ft [2]. These figures are well below the general requirement that the aircraft touchdown dispersion be below 40 ft longitudinally [5].

We also simulated the aircraft response to vertical burble shown in Fig. 9. Touchdown occurs at $t = 28$ sec, and the response of the aircraft is shown in Figs. 10 and 11. It can be seen from Fig. 10 that the decrease in airspeed of 1.2 ft/sec at touchdown is quite small. As shown in Fig. 11, the burble causes a 0.44-ft drop below the glide slope with the consequent touchdown about 8.4 ft short. This figure is also quite satisfactory. In comparison, with the H-dot control law, the aircraft touches down due to ship burble 8 ft long [2].

There are no specific requirements on the variance for the horizontal tail deflection in the presence of sensor noise. MIL-C-18244A states [6],

"Noise superimposed on a proper signal shall not saturate the automatic flight control system components and shall not cause objectionable motion of control stick or wheel."

For the control law given by (11), we now determine the variance of the horizontal tail deflection in the presence of the position sensor noise v_1. From (7), we have

$$\begin{pmatrix} \dot{x} \\ \dot{q} \end{pmatrix} = A_c \begin{pmatrix} x \\ q \end{pmatrix} + \bar{B}_c v_1, \tag{13}$$

where $\bar{B}_c$ is the second column of the matrix $\begin{pmatrix} B_2 \\ -LE_2 \end{pmatrix}$. Note that v_1 is the noise associated with measuring $\frac{h}{V}$ with the radar on board the ship. The radar senses the angular position of the aircraft. Hence, the radar noise is in units of radians. Thus v_1 is dependent on the distance of the aircraft from the carrier. For our purposes, it is sufficient to consider the situation where the aircraft is 0.5 miles from the carrier deck. The power spectral density of v_1 can be computed from AR-40A [5] as

$$\Phi_{v_1} = \frac{0.00256}{\left(\frac{j\omega}{50}+1\right)\left(\frac{-j\omega}{50}+1\right)}.$$

The transfer function from v_1 to δ_H can be easily computed from (11) and (13). The variance of the horizontal tail deflection due to v_1 is given by

$$\sigma^2_{\delta_H} = \int_{-\infty}^{\infty} \left|\frac{\delta_H}{v_1}(j\omega)\right|^2 \Phi_{v_1}\, d\omega.$$

The standard deviation σ_{δ_H} is computed to be 0.9 milliradians, which is quite small.

5. Conclusions

In this chapter we synthesized a control law for the Automatic Carrier Landing System of an F/A-18A using finite horizon H_∞ techniques. Utilizing output feedback, a suboptimal controller was derived that yielded good vertical rate command response. Experimentation with the suboptimal value and the weighting matrices was needed to obtain a satisfactory output feedback controller. In order to verify the performance of the controller on the complete aircraft, the longitudinal aircraft equations during the landing mode were augmented with the actuator and the engine dynamics. The resulting 17-dimensional model was tested using the low-order output feedback controller. The responses indicated that the low-order controller could adequately control the augmented aircraft model, with performance approaching close to that with the simpler longitudinal model. Also the response of the aircraft to vertical turbulence and ship burble was shown to be quite satisfactory.

References

[1] URNES, J. M., HESS, R. K., MOOMAW, R. F., AND HUFF, R. W., "Development of the Navy H-dot Automatic Carrier Landing System Designed to give Improved Approach Control in Air Turbulence," *Proceedings of the AIAA Guidance and Control Conference*, Boulder, Colorado, 1979, pp. 491–501.

[2] URNES, J. M. AND HESS, R. K., "Development of the F/A-18A Automatic Carrier Landing System," *Journal of Guidance, Control, and Dynamics*, Vol. 8, May–June 1985, pp. 289–295.

[3] "F/A-18A Flight Control System Design Report," Vols. I,II, and III, Report No. MDC A7813, McDonnell Aircraft Company, St. Louis, Missouri, September 1988.

[4] SUBRAHMANYAM, M. B., "General Formulae for Suboptimal H_∞ Control over a Finite Horizon," *International Journal of Control*, Vol. 57, No. 2, 1993, pp. 365–375.

[5] "Automatic Carrier Landing System, Airborne Subsystem, General Requirements for," AR-40A, Naval Air Systems Command, Washington, D.C., May 1975.

[6] "Control and Stabilization Systems: Automatic Piloted Aircraft, General Specifications for," MIL-C-18244A, Bureau of Naval Weapons, Department of the Navy, Washington, D.C., December 1962.

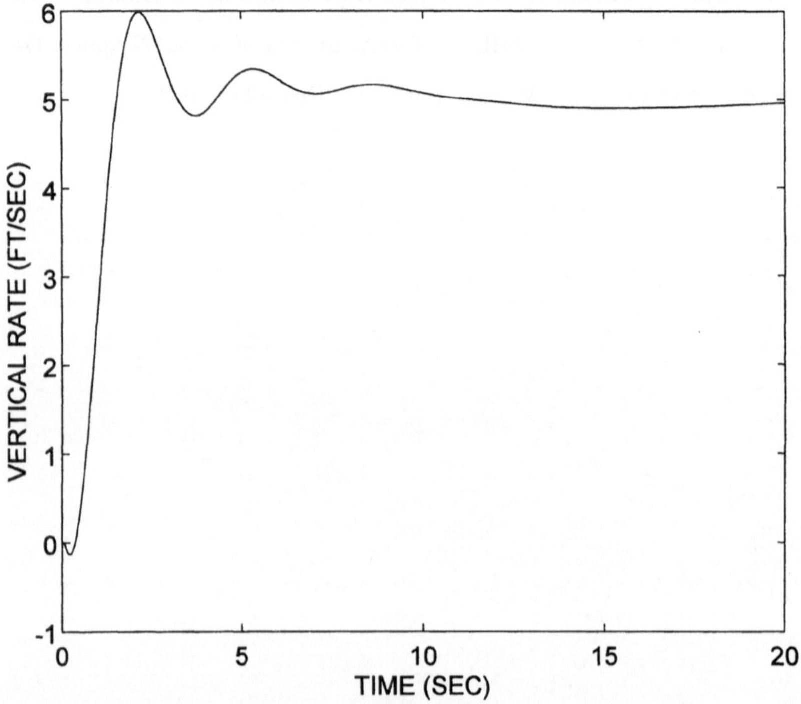

Fig. 1 Vertical rate response of the augmented plant

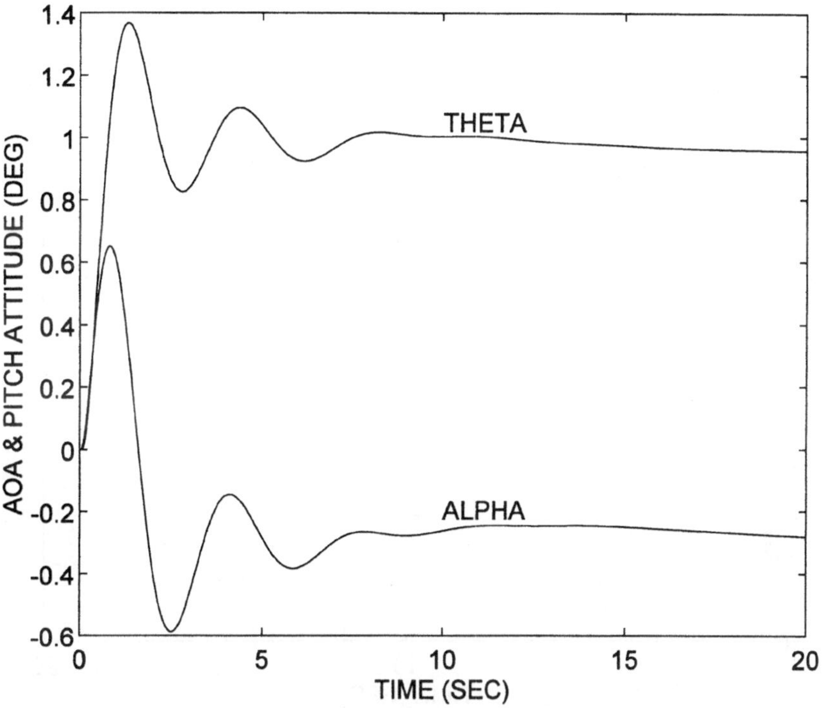

Fig. 2 Perturbed angle of attack and pitch attitude of the augmented plant

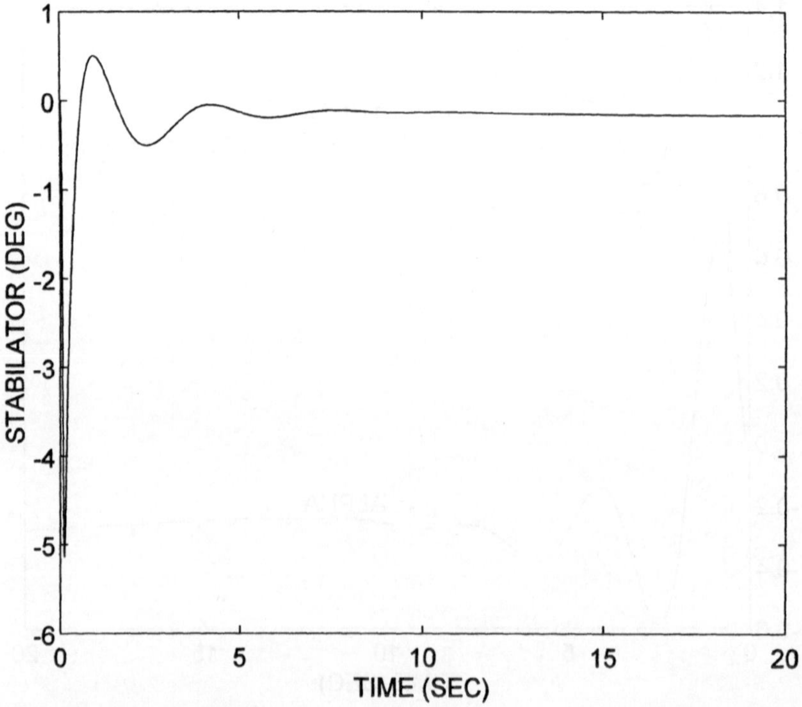

Fig. 3 Perturbed stabilator deflection of the augmented plant

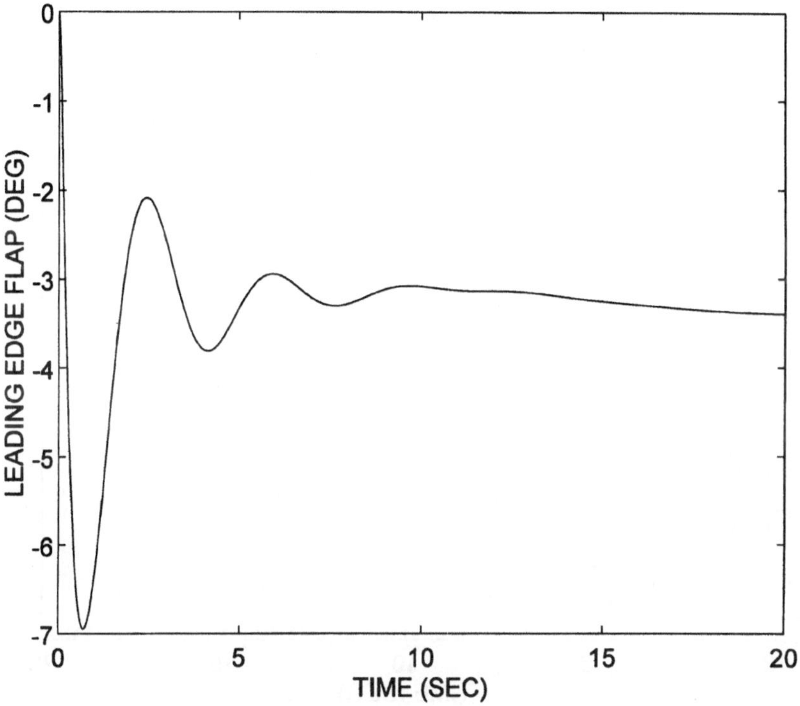

Fig. 4 Perturbed leading edge flap deflection

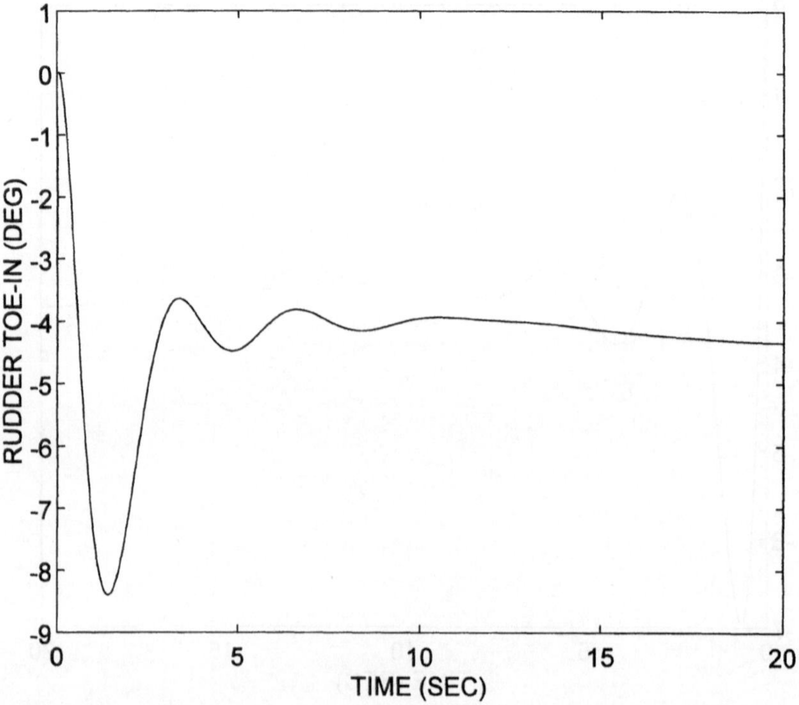

Fig. 5 Perturbed rudder toe-in

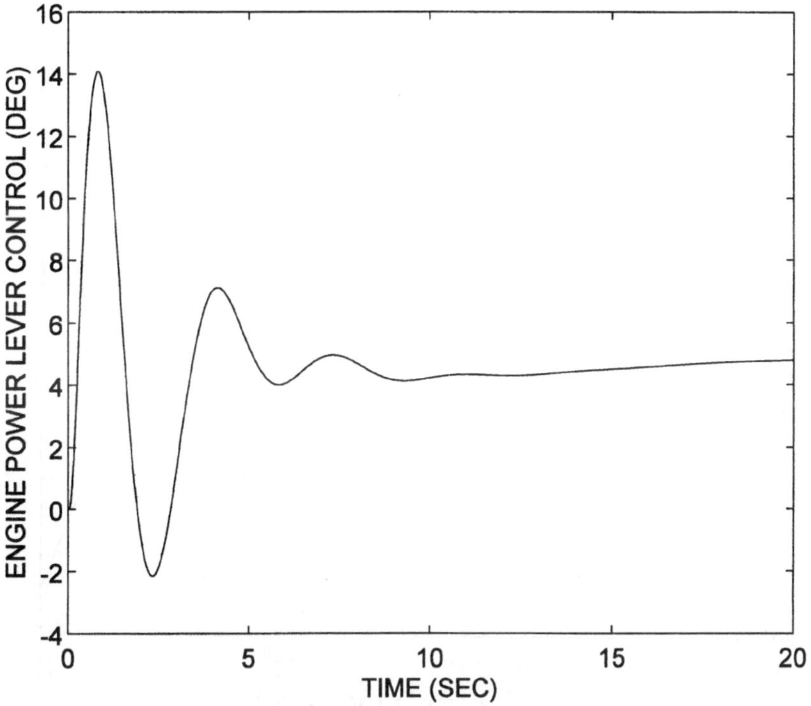

Fig. 6 Perturbed engine power lever control angle

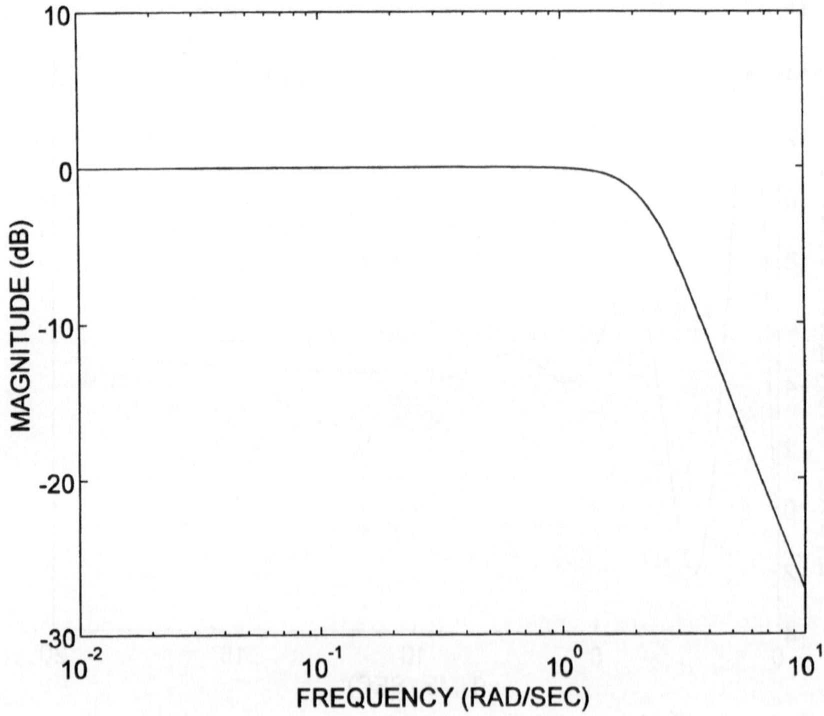

Fig. 7 Magnitude plot of vertical rate to vertical rate command

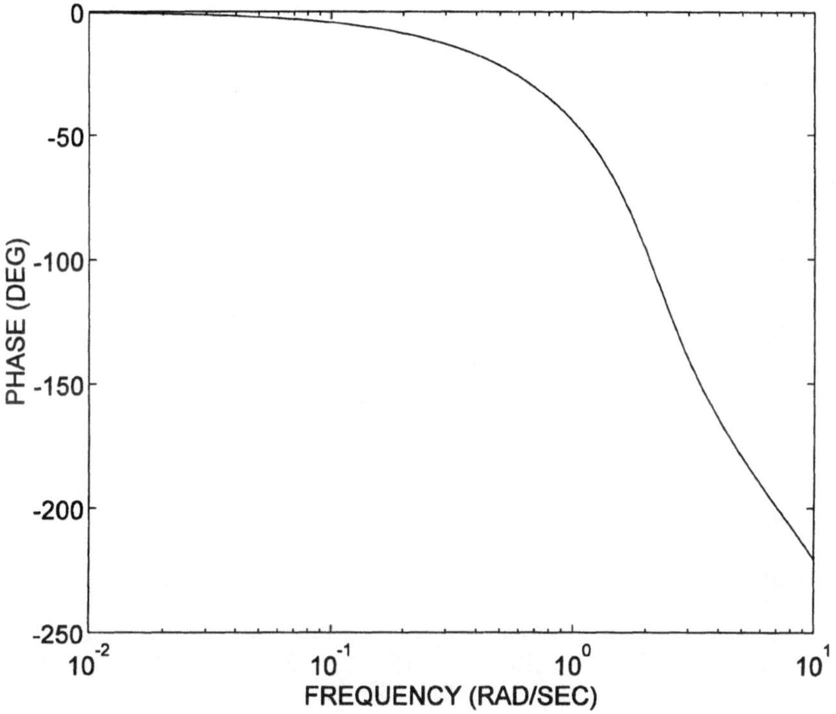

Fig. 8 Phase plot of vertical rate to vertical rate command

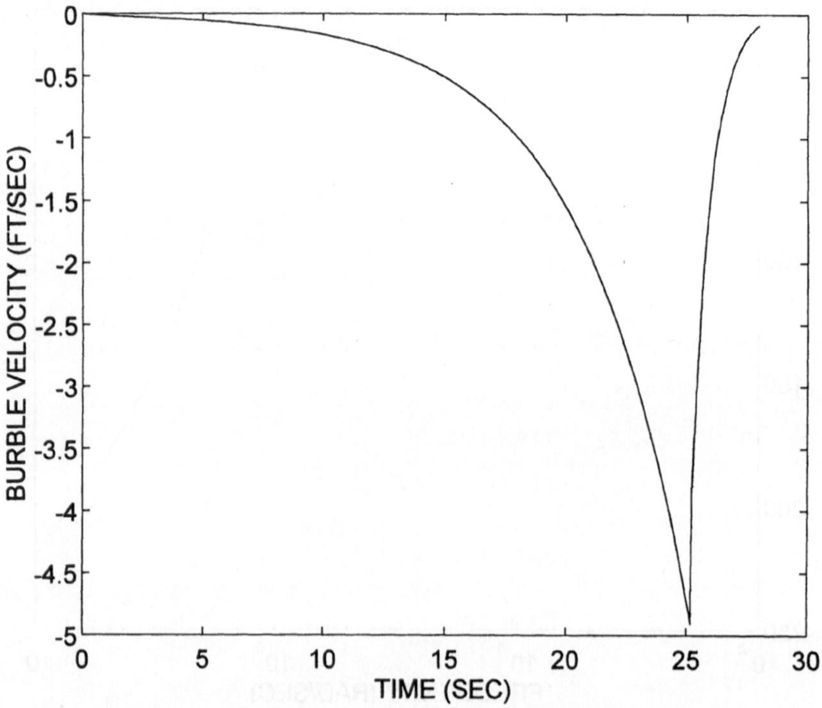

Fig. 9 Vertical burble velocity

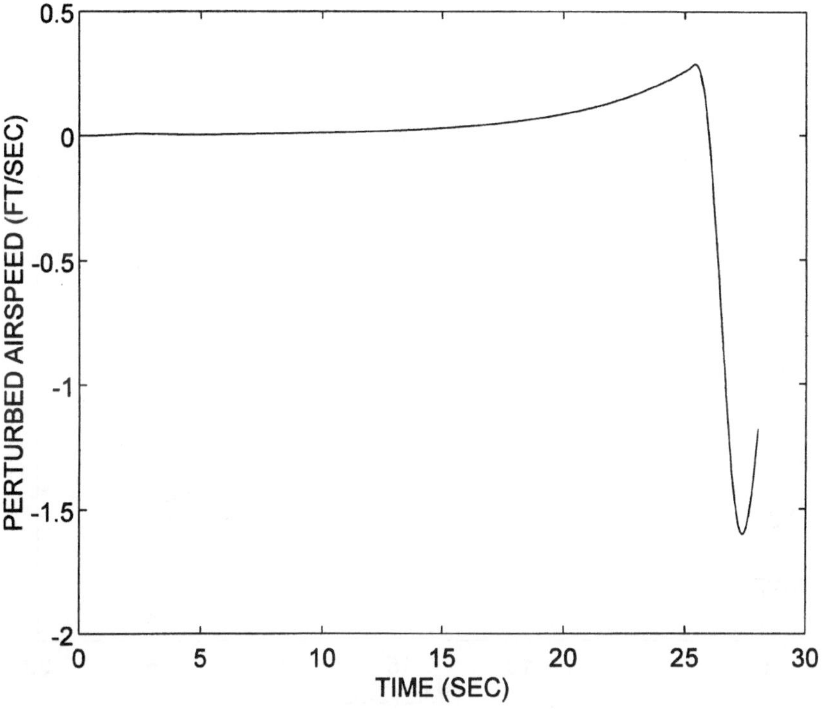

Fig. 10 Perturbed airspeed response due to ship burble

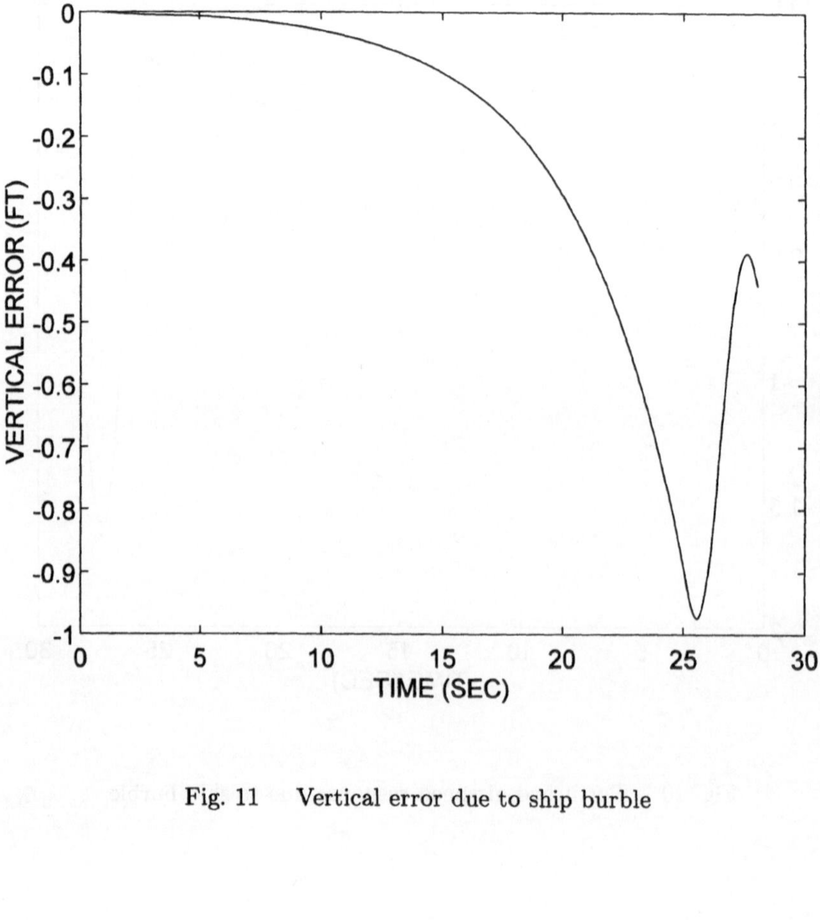

Fig. 11 Vertical error due to ship burble

Subject Index

actuator, 98
— dynamics, 98
adjoint
— response, 18,38,78
— variable, 38,45
— vector, 83,89
angle
— of attack, 94,95,97
bank —, 94
engine power lever control —, 95
pitch —, 95
Approach Power Compensation System (APCS), 93,100
— servo, 99
Automatic Carrier Landing System (ACLS), 93,103

Banach space, 2,75
bank angle, 94
boundary
— conditions, 80
— value problem, 20–23,29,31,61,62, 77,81,83,89

carrier air wake, 94
closed loop system, 30,73,74,88,90,98
condition
transversality —, 18,38
conditions
boundary —, 80
necessary —, 1,5,6,10,73,77,90
saddle point —, 30,35
sufficient —, 74
transversality —, 9
cone, 2,78
convex —, 2
dual —, 3,4,78
control
— feedback matrix, 97
— variable, 93
finite horizon H_∞ suboptimal — problem, 86
H_∞ —, 15
optimal —, 1,5,6
suboptimal H_∞ —, 35
controller, 49,57,66,68,70,73,86,88,90
feedback —, 15,16,29,32,42
finite horizon H_∞ —, 12
H_∞ —, 36,94
H_∞ — design, 94
H_∞ suboptimal —, 60
optimal —, 20,21
output feedback —, 16,35,36,42,50,94, 97,100,104
state feedback —, 16,28,31,37,39,58,88
suboptimal —, 35,36,50,53,54,58,70,88, 97,103
suboptimal finite horizon H_∞ —, 50,85
suboptimal H_∞ —, 1,36,37,53,54,66,73, 85
convergence
Lebesgue dominated — theorem, 76
weak —, 76
convex
— cone, 2
— functional, 16,35
cost functional, 1,10,11,73,78,90
criterion
performance —, 17,37

Deck Motion Compensation (DMC), 93
differential equation, 23,31
direction
— of decrease, 2,4,7,78
feasible —, 3,4,7
tangent —, 3,4,78,79

dual
— cone, 3,4,78
— space, 3,4
— system, 44
Dubovitskii-Milyutin theorem, 4,78,80
dynamics
actuator —, 98
engine —, 98,99,101
system —, 98

eigenvalues, 30,98,100
elemental variations, 83
engine, 99
— dynamics, 98,99,101
— power lever control angle, 95,101
equation
differential —, 23,31
Riccati —, 15,16,28,31,35,36,46,53,58, 59,88,93,97
error energy, 97
Euclidean norm, 75
existence theorem, 73
exogenous input, 29,30,40,61,62,89

F/A-18A, 93,94,96,103
feasible direction, 3,4,7
feedback
control — matrix, 97
full state —, 53
output —, 53,94,103
feedback controller, 15,16,29,32,42
state —, 16,28,31,37,39,58,88
output —, 16,35,36,42,50,94,97,100, 104
finite horizon
— H_∞, 53,73,93
— H_∞ controller, 12
— H_∞ norm, 22,25,73
— H_∞ optimal control, 15,30
— H_∞ performance robustness problem, 53,70,73,74,85,90
— H_∞ suboptimal control problem, 86
— suboptimal H_∞ problem, 36
suboptimal — H_∞ controller, 50,85
flight path angle, 93,94
forward velocity, 95
Fréchet derivative, 78,79
functional
convex —, 16,35
cost —, 1,10,11,73,78,90
linear —, 3
nonstandard —, 1
support —, 3

glide slope, 102

H_∞
finite horizon —, 53,73,93
finite horizon — norm, 22,25,73
finite horizon — performance robustness problem, 53,70,73,74,85,90
finite horizon — suboptimal control problem, 86
finite horizon suboptimal — problem, 36
— control, 15
— controller, 36,94
— controller design, 94
— design, 93
— norm, 74,90
— output feedback design, 101
— performance robustness problem, 74,82
— problem, 15,17,35
— suboptimal controller, 60
— suboptimal design, 94
infimal — norm, 15,25,31,53,94
suboptimal finite horizon — controller, 50,85
suboptimal — control, 35
suboptimal — controller, 1,36,37,53,54, 66,73,85
suboptimal — problem, 36,53,55
Hamiltonian, 18,38
horizontal tail, 95,102

identity matrix, 68,95
infimal H_∞ norm, 15,25,31,53,94
integral inequalities, 73

leading edge flap, 95,99,101
Lebesgue dominated convergence theorem, 76
linear
— functional, 3
— system, 1,73
— topological space, 2
Lipschitz condition, 2,11
longitudinal equations, 93
longitudinal system, 96
longitudinal velocity, 102

maximum principle, 18,38,60
matrix
control feedback —, 97
identity —, 68,95
transition —, 22,62
weighting —, 17
minimax
— problem, 17
— solution, 19
minimization problem, 73,74,90

necessary conditions, 1,5,6,10,73,77,90
nonlinear system, 1
nonstandard functional, 1
norm, 3
Euclidean —, 75
finite horizon H_∞ —, 22,25,73
H_∞ —, 74,90
infimal H_∞ —, 15,25,31,53,94

observer, 44,49,57,87
optimal control, 1,5,6
optimal controller, 20,21
optimal trajectory, 30,61

output feedback, 53,94,103
H_∞ — design, 101
— controller, 16,35,36,42,50,94,97, 100,104

parameter
— uncertainties, 53,54
— variations, 53,54,63,64,67,70,74, 82,85
performance
finite horizon H_∞ — robustness problem, 53,70,73,74,85,90
H_∞ — robustness problem, 74,82
— criterion, 17,37
— degradation, 54
— index, 36
— robustness, 54,73,85
— robustness problem, 54,66,67
suboptimal — robustness problem, 55,66, 90
— variation, 85
pitch
— angle, 95
— rate, 95
power spectral density, 102,103
principle
maximum —, 18,38,60
problem
boundary value —, 20–23,29,31,61, 62,77,81,83,89
finite horizon H_∞ performance robustness —, 53,70,73,74,85,90
finite horizon H_∞ suboptimal control —, 86
H_∞ —, 15,17,35
H_∞ performance robustness —, 74,82
minimax —, 17
minimization —, 73,74,90
performance robustness —, 54,66,67
suboptimal H_∞ —, 36,53,55
suboptimal performance robustness —, 55,66,90

radar, 93,103
response
adjoint —, 18,38,78

Riccati equation, 15,16,28,31,35,36,46,53,58,59,88,93,97
robustness
finite horizon H_∞ performance — problem, 53,70,73,74,85,90
H_∞ performance — problem, 74,82
performance —, 54,73,85
performance — problem, 54,66,67
suboptimal performance — problem, 55,66,90
rudder, 99
— toe-in, 95,101

saddle point, 18,37
— conditions, 30,35
— solution, 19,39,45
sensor noise, 93,94,102
ship burble, 93,102,104
sink rate, 102
smooth manifolds, 9
solution
minimax —, 19
saddle point —, 19,39,45
space
Banach —, 2,75
dual —, 3,4
linear topological —, 2
stabilator, 99,101
state
full — feedback, 53
— feedback controller, 16,28,31,37,39,58,88
suboptimal controller, 35,36,50,53,54,58,70,88,97,103
suboptimal finite horizon H_∞ controller, 50,85
suboptimal H_∞ control, 35
suboptimal H_∞ controller, 1,36,37,53,54,66,73,85
suboptimal H_∞ problem, 36,53,55
suboptimal performance robustness problem, 55,66,90
sufficient conditions, 74
support functional, 3
supporting hyperplane, 3
system
closed loop —, 30,73,74,88,90,98
dual —, 44
linear —, 1,73
longitudinal —, 96
— dynamics, 98
— matrices, 53,54,63,66,70,73,85,90,95
time-invariant —, 53,59,88
time-varying —, 16,27,36,55

tangent direction, 3,4,78,79
theorem
Dubovitskii-Milyutin —, 4,78,80
existence —, 73
Lebesgue dominated convergence —, 76
thrust, 93
time-invariant system, 53,59,88
time-varying system, 16,27,36,55
transfer function, 97,102,103
transition matrix, 22,62
transversality
— condition, 18,38
— conditions, 9
turbulence, 101,104

variable
adjoint —, 38,45
variation
performance —, 85
variations
elemental —, 83
parameter —, 53,54,63,64,67,70,74,82,85
vector
adjoint —, 83,89
— transpose, 17,37,55,95
vertical burble, 101,102
vertical gust, 93,94,95,101
vertical rate command, 93,97,101

weak convergence, 76
weighting matrix, 17
worst disturbance, 19

Systems & Control: Foundations & Applications

Series Editor
Christopher I. Byrnes
School of Engineering and Applied Science
Washington University
Campus P.O. 1040
One Brookings Drive
St. Louis, MO 63130-4899
U.S.A.

Systems & Control: Foundations & Applications publishes research monographs and advanced graduate texts dealing with areas of current research in all areas of systems and control theory and its applications to a wide variety of scientific disciplines.

We encourage the preparation of manuscripts in TEX, preferably in Plain or AMS TEX LaTeX is also acceptable—for delivery as camera-ready hard copy which leads to rapid publication, or on a diskette that can interface with laser printers or typesetters.

Proposals should be sent directly to the editor or to: Birkhäuser Boston, 675 Massachusetts Avenue, Cambridge, MA 02139, U.S.A.

Estimation Techniques for Distributed Parameter Systems
H.T. Banks and K. Kunisch

Set-Valued Analysis
Jean-Pierre Aubin and Hélène Frankowska

Weak Convergence Methods and Singularly Perturbed Stochastic Control and Filtering Problems
Harold J. Kushner

Methods of Algebraic Geometry in Control Theory: Part I
Scalar Linear Systems and Affine Algebraic Geometry
Peter Falb

H^∞-Optimal Control and Related Minimax Design Problems
Tamer Başar and Pierre Bernhard

Identification and Stochastic Adaptive Control
Han-Fu Chen and Lei Guo

Viability Theory
Jean-Pierre Aubin

Representation and Control of Infinite Dimensional Systems, Vol. I
A. Bensoussan, G. Da Prato, M. C. Delfour and S. K. Mitter

Representation and Control of Infinite Dimensional Systems, Vol. II
A. Bensoussan, G. Da Prato, M. C. Delfour and S. K. Mitter

Mathematical Control Theory: An Introduction
Jerzy Zabczyk

H_∞-Control for Distributed Parameter Systems: A State-Space Approach
Bert van Keulen

Disease Dynamics
Alexander Asachenkov, Guri Marchuk, Ronald Mohler, Serge Zuev

Theory of Chattering Control with Applications to Astronautics, Robotics, Economics, and Engineering
Michail I. Zelikin and Vladimir F. Borisov

Modeling, Analysis and Control of Dynamic Elastic Multi-Link Structures
J. E. Lagnese, Günter Leugering, E. J. P. G. Schmidt

First Order Representations of Linear Systems
Margreet Kuijper

Hierarchical Decision Making in Stochastic Manufacturing Systems
Suresh P. Sethi and Qing Zhang

Optimal Control Theory for Infinite Dimensional Systems
Xunjing Li and Jiongmin Yong

Generalized Solutions of First-Order PDEs: The Dynamical Optimization Process
Andreĭ I. Subbotin

Finite Horizon H_∞ and Related Control Problems
M. Bala Subrahmanyam